现代轻奢

MODERN LIGHT LUXURY

当代住宅设计
Contemporary Residential Design

/ 深圳视界文化传播有限公司 编 /

中国林业出版社
China Forestry Publishing House

PREFACE

定制化轻奢设计，阐述时代沉淀的品位

早在上世纪 90 年代，全球范围内便存在一种生活风尚：低调奢华 (Low Profile Luxury)。这是一种体现优雅品位、务实品质的生活方式，淡化物质追求，转而开始注重个人审美熏陶和精神世界的塑造。例如，艺术品收藏、攀岩、划艇甚至沙漠摄影等等，强调不张扬、不随俗却严谨、克制而个性化，充分表达了人们对高品质生活的热爱和追求、对生活美学的探索与享受。时至今日，这种生活哲学仍然经久不衰，甚至愈演愈烈，社会精英、金融才俊、企业领袖等皆以这样的生活追求而感到自豪！

设计界同样受到这种"轻奢"理念的深刻影响。时装界以奢侈品 Armani 为首，创造时尚品牌轻奢副线，而家具界则以 Giorgetti 为首，作品设计集实用性与艺术性于一身，创造独特、折衷的风格品牌。

2010 年以后，中国室内设计界逐渐兴起一股"轻奢风"，然而专业书籍中，并没有一本针对"轻奢"设计作全方位案例汇总的工具书出现。这次《现代轻奢》这本集大成之作面世，有幸先睹为快。书中分为大陆、港式和台式轻奢三大板块，以"轻奢华，新时尚"为全书主旨，集结了两岸三地众多轻奢风格设计作品，图文并茂的内容编排给人扎实的阅读体验。

面对日益提升的审美品位，室内装饰也逐渐转向独特的定制化轻奢设计，设计风格因地方差异而形成不同风味：大陆轻奢前卫大方、港式轻奢时尚商务、台式轻奢沉稳精致。因地制宜的设计各有千秋，其设计细节的奥妙之处在于：轻——黑白灰主调与原木材质的穿插交替，赋予住宅时尚轻盈的空间气氛；奢——香槟金不锈钢或黑镜钢的运用，以简约的线条勾画空间气质，让空间轮廓笔直而清晰、雅致且高傲。

眨眼已在室内设计这个领域走过 33 个年头，以经验论道，我将"轻奢"理解为"Mild Luxury"——温和而适度，淡雅而高贵，它的气质表现可概括为：

以温柔见高贵

以干净见时尚

以低调见个性

以纯色见品位

我相信，如果您读懂这本《现代轻奢》，必定会为您带来一场设计的头脑风暴，同时让您对当代设计的脉搏有更清晰的把握！在此非常感谢总编邀写序言，幸甚！最后送诸君一句话：设计无大小，用心做的就是好设计！

设计师 / 李奇恩 Henry Li
现任尚策设计顾问有限公司创办人及董事长
深圳室内设计师协会（SZAID）副秘书长
香港室内设计协会（HKIDA）专业会员

1985 年从事室内设计行业，葆有一颗赤子之心，天性乐观开朗，具有娱乐精神，自称"老顽童设计师"。参与设计行业之发展。2009 年创立尚策设计，带领团队先后新获国内外众多设计大奖，其中不乏德国 Red Dot Aware、意大利 A' Design Award&Competition、新加坡 Singapore Interior Design Awards、德国 ICONIC Awards 等国际性奖项，并屡受媒体邀请担任演讲嘉宾及相关活动主持人。

CUSTOMISE THE LIGHT LUXURY DESIGN AND STATE THE TASTE OF TIME PRECIPITATION

Since the 1990s, there is a lifestyle around the world, that is the Low Profile Luxury, which represents the elegant tastes, the quality lifestyle and fading out the pursuit of materials. It encouraged people to pay attention to edifying the individual aesthetic and building spirit worlds, such as the collection of work of art, climbing, canoeing and taking photos in the desert, which emphasised that people should be restraining and nonconformity, and they should personalise their life with preciseness and restraint. It fully expressed the devotion and pursuit people have for their high-quality life and the exploration and enjoyment for the aesthetic of life. Even to this day, this philosophy of living remains, and more and more people, such as the world elites, the talented financial people and business leaders, are proud of living like this.

This style also influences design. In the fashion world, the luxury brand, Armani, for example, creates a side-line product of light luxury and in the furniture world, Giorgetti for instance, it designs the products with practicability and artistry which builds a brand with a style of unique and compromised.

After 2010, the interior design in China has risen a style of light luxury. However, there is not a reference book combines all kinds of the light luxury design. Modern Light Luxury is the first one to collect all these designs together and publish them to audiences. This book can be divided into three parts, light luxury style in Mainland, light luxury style in HongKong, light luxury style in Taiwan. The gist of this book is "Light Luxury and New Vogue", and it collects many light luxury designs from three places of two sides with the picture and its accompanying essay to allow audiences to understand the light luxury design.

Because of the gradually enhanced the aesthetic tastes, the interior decorations have changed direction into the unique customised light luxury design. The various styles of design are based on different place and have a unique flavour. The light luxury in the mainland is generous, while in Hong Kong, it is fashion business, and in Taiwan, it is steady-going and exquisite. Each design has its own merits, and the secrets of the details are light and luxury. For the light factor, the designer uses black, white and grey tones with wood materials to give the house a sense of lightness and fashion. For the luxury factor, the designer uses champaign gold stainless steel and jet black steel to outline of space with simple lines which shows the straight and clear, elegant and arrogant of the space profile.

Now, the interior design has developed for 33 years, and I will express the definition of "light luxury" with the experiences that are mild and moderate, simple and elegant. The temperament can be described as:

Gentle and elegant

Clean and fashion

Low profile with characteristic

Pure colour with taste

We hope after finishing this book, it can bring you brainstorm and allow you to have a clearer understanding of modern design. Thanks so much the editor-in-chief invites me to write this forward and in the end, I would like to tell people that there is no big or small design in the world, the best is with the designer's hard work.

目/录 CONTENTS

006 CHAPTER ONE
大陆轻奢 LIGHT LUXURY STYLE IN MAINLAND

- *010* 简致灰调层染本质生活
  Simple Grey Tone Dyes the Essence of Life

- *028* 岁月如歌，依山而居
  Time Is Like A Song and We Live Rely On Mountains

- *040* 黑白秩序下的现代奢雅
  The Modern Luxury and Elegance Based On The Order of Black and White

- *054* 品质生活，流连于繁华之后
  Enjoying the Quality Life After Bustling

- *066* 流动在白色空间里的艺术乐章
  Art Movement Flowing in White Space

- *082* 综艺导演的艺术栖所
  The Artist Place for Variety Show Director

- *098* 浸润于轻奢，陶冶出优雅
  Soaking into Light Luxury and Cultivating Elegance

- *108* 极致顶复的简雅精神
  The Simplicity and Elegant Spirit of The Extreme Penthouse

- *120* 坐拥四季的清雅·冬
  Enjoying the Elegant Four Seasons · Winter

128 CHAPTER TWO
港式轻奢 LIGHT LUXURY STYLE IN HONGKONG

132 不言奢华，只言生活
Not Talking About Luxury, We Only Talk About Life

146 轻质设计 高感时尚
Light Design, High Fashion

162 以传统设计章法打造现代住宅
Using Traditional Design to Build Modern House

172 繁华之上，浮沉随心
On the Bustling Life and Having A Beautiful House

180 精致，简奢于心
Delicacy, Simplicity and Luxury In Heart

198 褪去浮华 舒享于简
The Faded Glitz Return to the Nature, and Feeling the Cozy and Simple Life

206 简致设计合围人生清欢
Simple and Exquisite Design which Making Faint Joy and Tranquil Life

220 身心栖居的"世外桃源"
Living in A Land of Idyllic Beauty

232 方圆之间，奢于亲人陪伴
To Company Your Families as Luxurious as Possible in Fangyuan

CHAPTER THREE 240
LIGHT LUXURY STYLE IN TAIWAN 台式轻奢

热带住宅的恢弘与度假情调 *244*
The Grand Tropical House and Vacation Sentiment

退休企业家的魅力宅邸 *260*
The Amazing House of Retired Enterpriser

华丽与抽象的演绎 *270*
The Performance of Gorgeous and Abstraction

格调家居演绎生活序曲 *284*
Style Furniture Performs Prelude to Life

城市逸墅，迎光筑景 *294*
Urban Villa, Building Scenery to Meet the Sunlight

筑有温度的低奢场域 *310*
The Dense Inked Flavour Building A Space with Temperature

LIGHT LUXURY
MAINLAND

现代轻奢 / Modern Light Luxury

STYLE IN

大陆轻奢
LIGHT LUXURY STYLE IN MAINLAND

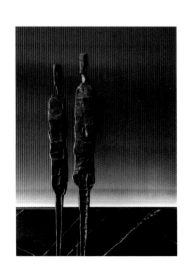

轻奢理念 / The Concept of Light Luxury
轻奢聚焦 / The Focus of Light Luxury
轻奢生活 / The Life of Light Luxury

轻奢印象
THE IMPRESSION OF LIGHT LUXURY

大陆轻奢设计以现代主义风格为主旋律，集极简主义的简洁明朗与装饰主义"合乎其功利要求的美化"于一身，追求时尚与潮流，注重空间布局与使用功能的完美统一，设计中常掺杂后现代风格的混搭手法，增加室内空间形态的单一性与抽象性。

Light luxury style in Mainland is mainly focused on the modernist. It combines the simplicity of minimalism with decoration, which meets the beautification of functional requirements. It pursues fashion and trend, pays attention to the perfect unity of spatial layout and function. The mix of post-modern style increases the unicity and abstraction of the interior space.

1　DIA 丹健国际 ｜ 深圳汉京九容台
　　客厅

2　Ippolito Fleitz Group ｜ 上海中鹰·黑森林
　　餐厅一角

3 DIA 丹健国际丨深圳汉京九容台
　客厅

4 余颢凌丨白麓
　楼梯区

01 轻奢理念　THE CONCEPT OF LIGHT LUXURY

　　轻奢设计重视空间的实用性、灵活性和高使用率。注重会客、餐饮、学习、睡眠等功能空间的逻辑关系，即空间关系不再是单纯的房间组合，而是根据各空间的功能关系，相互配合、相互渗透而成。空间划分不再局限于实体墙，而是通过天花板、照明、家具布置、地坪等的变化来进行区别。此外，充分利用时间差，为同一空间的不同时间段打造不同角色，充分表现空间的灵活性、兼容性和流动性。

The light luxury design emphasises the practicality, flexibility and high usage of space. Focusing on the logical relationship of functional space, such as meeting, dining, learning and sleeping, that is to say, the spatial relationship is no longer a simple combination of rooms, but based on the functional relationship of each space, cooperation and penetration. Space division is no longer limited to solid walls but is differentiated by changes in ceilings, lighting, furniture layouts and floors. Also, it makes good use of the time difference in creating different roles for different periods in the same space, and fully demonstrating the flexibility, compatibility and mobility of the space.

02 轻奢色彩　THE COLOUR OF LIGHT LUXURY

　　配色灵活而丰富，通常在现代简约的基础上进行纯净色调的相互搭配，增加室内金属色、香槟色等，着重打造轻奢舒适、个性化的空间。

The colour matching is flexible and abundant. Usually, the pure colour is matched with each other from the modern simplicity and adding the colour indoors such as metallic and champagne, which makes a light luxury comfort and personalised space.

03 轻奢用材　THE MATERIAL OF LIGHT LUXURY

　　采用新技术和新材料，注重环保，重视材料的质感与性能，在强调功能性的前提下突出设计的时尚气质。应用金属、漆面、钢化玻璃等新型材料为空间创造前卫、不受拘束的感觉；皮毛、水晶、贝壳等天然材料增加住宅的自然轻奢品质。

It adopts new technologies and materials, pays attention to environmental protection, texture and performance of materials, highlights the elegant temperament of design under the premise of emphasising functionality. The use of new materials, such as metal, lacquer surface and tempered glass, creates an avant-garde and unconstrained feeling for space. Natural materials, such as fur, crystal and shells increase the natural luxury quality of the house.

04 家具配饰　FURNITURE ACCESSORIES

　　家具设计强调功能性——"形式服从于功能"，减少附加装饰，淡化物质追求，满足居者精神需求；推崇科学合理的构造工艺，家具造型简约、线条流畅，重视发挥材料自身的性能和特点。

It emphasises the functional design -- "forms are subject to function" additionally reduces decorations, dilutes material pursuit and meets the spiritual needs of the residents. It promotes the scientific and rational construction techniques, simple shape, smooth lines, develops performance and characteristics of the material itself.

SIMPLE GREY TONE DYES THE ESSENCE OF LIFE

简致灰调层染本质生活

LIGHT LUXURY STYLE IN MAINLAND

大陆轻奢

● 项目信息　PROJECT INFORMATION

项目名称 / 上海中鹰·黑森林
设计公司 / Ippolito Fleitz Group
项目地点 / 上海
项目面积 / 250 m²
摄影师 / Sui Sicong

扫码查看电子版

This Spatial Design uses various levels and stability in grey tones such as dark grey, light grey, bright grey and dim grey to create spatial levels and at the same time, avoids pressure and tiredness. Different textures and lines come together to develop abundant rhythm, and the colour and texture show the theme through "grey dye level". The level produced by the colour and texture adds more exploratory for this space and builds a perfect house for nowhere to place soul with the beautiful skyline.

THE CONCEPT OF LIGHT LUXURY
轻奢理念

空间设计运用了深灰、浅灰、明灰、暗灰等不同层次、不同饱和度的灰色调，灰色的明暗变化创造丰富的空间层次感，避免了高彩度色彩给视觉上带来的压迫感和疲惫感；不同质感、不同纹理的织物交织在一起协奏出丰富的空间节奏，色彩与材质演绎"灰色层染"的设计主题。色彩与材质塑造的层次感为空间增添了更多的探索性，与窗外优美的天际线一起，为城市里无处安放的灵魂和情感构筑一个完美归宿。

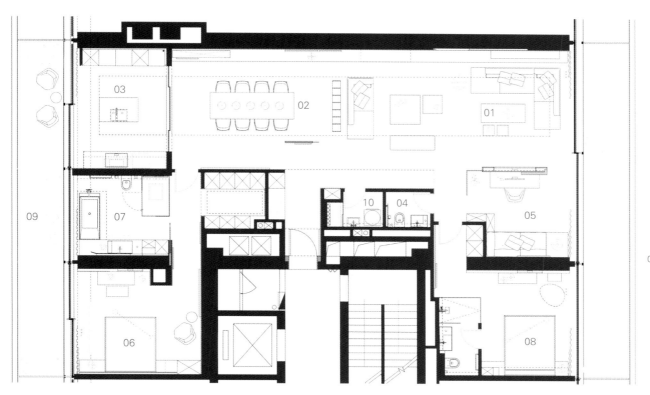

平面图 / SPACE PLAN

01 客厅 / Living Room
02 餐厅 / Dining Room
03 厨房 / Kitchen
04 次卫 / Secondary Bathroom
05 书房 / Study Room
06 主卧 / Master Bedroom
07 主卫 / Master Bathroom
08 次卧 / Secondary Room
09 阳台 / Balcony
10 客卫 / Guest Bathroom

THE FOCUS OF LIGHT LUXURY
轻奢聚焦

艺术品渲染空间气质，设计格局延伸住宅意象。踏入门厅即是一幅墨绿色的艺术挂画，宁静舒缓；居室内各个功能区的开放式布局让空间有机串联，不受阻挡的视线游移能给人宽敞的空间感。

彩色铝板天花与简约时尚的吊灯相结合，为包括起居室和厨房在内的整个休息、用餐区创造一个舒适温馨的氛围。

空间里点缀的绿色和蓝色装饰品与纯色地板形成强烈对比，成为设计亮点，使得整个空间变得更加现代、优雅与稳重。设计师善于运用材质与颜色上的混合拼接营造空间层次感，这还体现在起居室的设计里——绒面地毯、光滑的大理石和镜面玻璃相结合；大型家具与香槟色饰品相结合等，都为空间增添了时尚气质、轻质奢华感。

温暖的灰色基调、天然木质表面、雅致的灯光设计以及华丽的地毯与卧室中光滑的墙面形成了材质对比，营造出轻松的居室氛围。生动的大理石纹理、白色的卫浴产品以及香槟色与黑色组合的配饰完美结合，在浴室展现出它古典而永恒的清雅气质。

1 地板人字形铺设和线性天花板创造延伸的视觉效果与流畅的空间感受
2 宁静温馨的卧室配色
3 精致的材料切割与收口

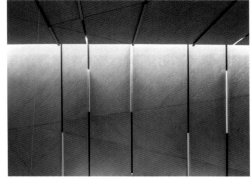

THE LIFE OF
LIGHT LUXURY
轻奢生活

轻奢是物质与精神相结合的生活态度，材质侧重表达物质，精致做工则更注重精神。一处处开窗纳入自然风景，顺室内门窗外延的阳台充满生机，劈面石材矮墙、窗户式样护栏、绿植连接无边天际，室内外达到自然、美学与机能一体的效果，召唤出一种感性、优雅的韵味，满足真正可贵的生活诉求。

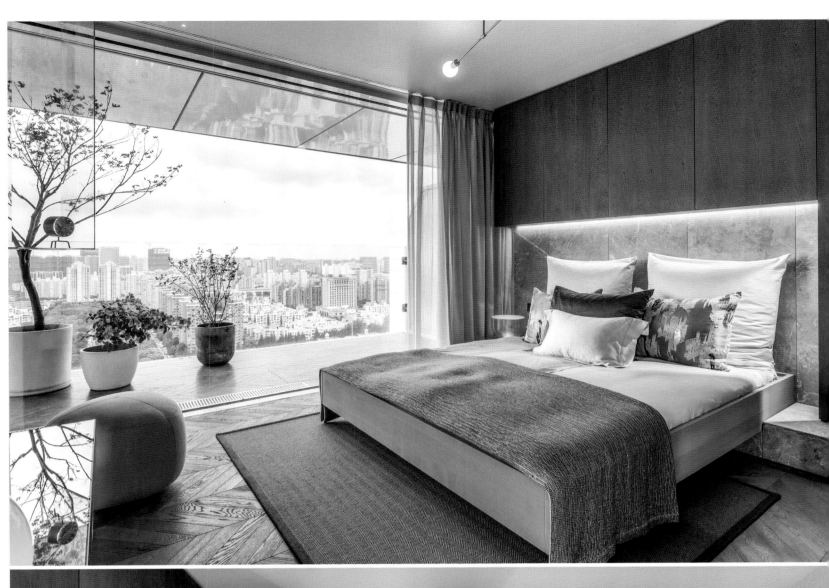

TIME IS LIKE A SONG AND WE LIVE RELY ON MOUNTAINS

岁月如歌,依山而居

LIGHT LUXURY STYLE IN MAINLAND
大陆轻奢

● 项目信息　PROJECT INFORMATION

项目名称 / 深圳汉京九容台 顶复
设计公司 / DIA 丹健国际
主创设计师 / 薛峰、张健、Manfred Haverkamp
设计团队 / 徐传鹏、王君明、张卫、桂菁儿
项目地点 / 广东深圳
摄影师 / 罗文

扫码查看电子版

THE CONCEPT OF LIGHT LUXURY

轻奢理念

If a person chooses to live in mountains and hides all his ambitious, he will return to the originals by washing all the attachments. The designer uses a "less is more" way to design this quiet and elegant house without piling up anything else.

　　邻山而居，隐心于野，大多会有洗尽铅华而回归本源的心境，为了塑造淡泊致远却又高级典雅的氛围，设计师在手法上很是克制，空间内未有丝毫堆砌，少即是多，点到为止。

THE FOCUS OF LIGHT LUXURY
轻奢聚焦

住宅坐落于"山丘"之中，独享300,000 m²天然绿肺，整体错落有序的建筑布局、丰富多变的视觉景象，犹如梦幻水晶与山体无边界的融合，居于其中，品味岁月给的馈赠，享受攀越巅峰后的宁静。

通高起居室连接正对大海的超大露台，业主坐拥俯瞰山麓的最佳视野，也独享静谧的专属空间。

克制意味着极致设计，但不冷漠，既然是居所，就要有家的温馨感，暖棕色穿插在黑白灰之间，充满生机的绿色与窗外的葱郁呼应，使人无时无刻都感受到恬淡舒适。

岁月会给人生带来什么？李宗盛用十年光阴写一首《山丘》来解答——是"不自量力去还手，至死方休"的勇气，是"嬉皮笑脸，面对人生"的乐观，还是"越过山丘"，俯瞰来时路的胸怀与赤胆？

1 天然石材切割的长桌

2 现代风格家具：造型简约，线条流畅，回应居者对生活的高品质追求

3 楼梯木质扶手：追求造型的同时精工材料接口，让每一根线条都流畅舒适

一层平面图
1st floor plan

01 客厅 / Living Room
02 餐厅 / Dining Room
03 西厨 / Western Kitchen
04 中厨 / Chinese Kitchen
05 阳台 / Balcony
06 客卫 / Guest Bathroom
07 影音室 / Video Room
08 藏酒室 & 雪茄室 / WINE & GIGAR
09 工人房 / Worker's Room

二层平面图
2nd floor plan

01 主卧 / Master Bedroom
02 主卫 / Master Bathroom
03 衣帽间 / Walk-in Closet
04 卧室 A / Bedroom A
05 卧室 B / Bedroom B
06 卫生间 / Bathroom
07 电梯厅 / Elevator Lobby
08 中空 / Hollow
09 过道 / Corridor

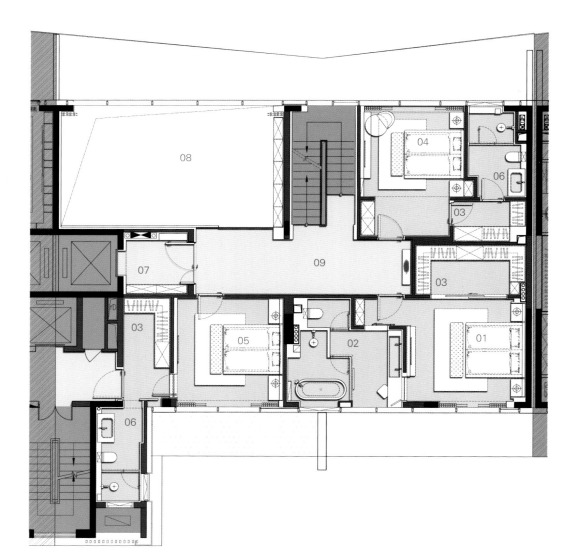

THE LIFE OF LIGHT LUXURY
轻奢生活

　　正如看尽浮华的人生，无需过多遮掩与点缀。隐居山丘之上，一览他山之小，这是每个成功人士的渴求，也是这套顶层复式不可复制的体验。越过山丘，邂逅极致居所；越过山丘，借山而居，那里有家，还有梦。

THE MODERN LUXURY AND ELEGANCE BASED ON THE ORDER OF BLACK AND WHITE

黑白秩序下的现代奢雅

大陆轻奢 LIGHT LUXURY STYLE IN MAINLAND

● 项目信息　PROJECT INFORMATION

项目名称 / 成都博瑞优品道黄龙溪谷 F2 户型　　项目地点 / 四川成都

设计公司 / 元禾大千　　项目面积 / 530 m²

轻奢理念

THE CONCEPT OF LIGHT LUXURY

Huanglongxigu is located at the top of ridges which has almost 100 lake acres. At Huanglong Valley, people can see the glistening blue light of waves, the original ecology valley and beautiful views. The designer, which follows the Modernism of Bauhaus School (started at the early stage of the 20th century) pursues simplify with pleasure. Also, the designer plays the colour games to the owner to have a warm feeling and reflect the hidden elegant, gorgeous and fashion of the space .

黄龙溪谷于山脊之巅，揽溪谷近百亩湖域，铺陈满目的蔚蓝波光，坐望原生大峡谷，将溪谷好景观纳入眼中。设计师从源于20世纪初包豪斯学派的现代主义出发，追求去繁化简但不寡淡无味，并在此基础上，玩转色彩游戏，让居者感受家的温度，反映空间隐藏着的高贵、优雅和时尚。

THE FOCUS OF LIGHT LUXURY
轻奢聚焦

色彩搭配，择金饰于经典黑白间

　　黑白色是本案的视觉担当，给人以沉稳、内敛的第一感受，同时以米色、灰色配合主色调，逐渐显现现代主义的主旋律。黑白色系往往给人以冰冷、凌冽之感，于是设计师选择金色饰于其中，从边几上的摆件，到空间内输出的暖色光源，无一不体现着设计师的细致缜密。

将家具融入空间

　　不少人认为，流线型家具过于充满艺术感，难以融于家居空间之中，但相比硬朗的直线型家具，圆融形态实则更能营造祥和的家庭氛围。而色彩同样选择空间主要用色，运用饱和度不一的搭配，丰富空间层次感。这正如设计师所说，现代简约设计更体现在概念生成阶段，而非在形式上，在设计的过程中将想法层层剥落，露出其坚硬的内核，最后产生出极美妙的效果。而这种极致的简约，就是一种极致的低调奢华。

→ 艺术灯沿用家人共享空间所用灯具的圆环光圈造型与元素，丰富卫浴空间的视觉呈现效果。

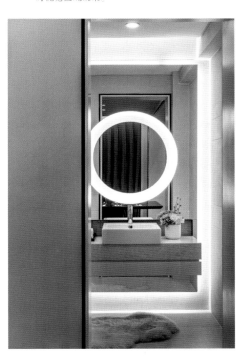

THE LIFE OF LIGHT LUXURY
轻奢生活

简约是现代主义的要点之一，其以将空间中的色彩、照明、材质简化到最少的程度为基础法则。设计师以此为参考，精准把握色彩与家具的选用，使家具看似简洁朴素，实则隐含奢雅，颇具"小隐隐于野，大隐隐于市"的哲学意味。

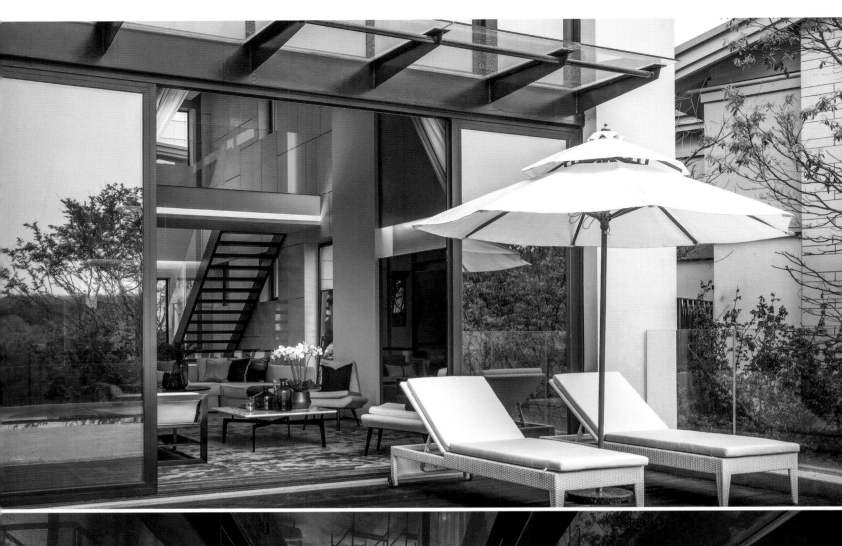

ENJOYING THE QUALITY LIFE AFTER BUSTLING

品质生活，流连于繁华之后

大陆轻奢 LIGHT LUXURY STYLE IN MAINLAND

● 项目信息　PROJECT INFORMATION

项目名称 / 杭州万科·大都会 79 号空中顶跃
设计公司 / MDO 木君建筑设计咨询（上海）有限公司
设计师 / 徐仗君、桥义
项目地点 / 浙江杭州

项目面积 / 406 m²
主要材料 / 石材（奥斯卡灰、意大利鱼肚白、葡萄牙木纹石等）、胡桃木饰面、胡桃木地板、白杨木饰面、皮革、壁纸、地毯、布艺等
摄影师 / 光影摄影（万科授权）

扫码查看电子版

THE CONCEPT OF LIGHT LUXURY
轻奢理念

The designer believes that design can bring the power of "change", and it changes the bland environment into the special spatial experience and leads people from one mood into another. The design aims to guide people to a "Change" journey, and the designer uses the light, the combination of different sizes and the combination of different materials to achieve this goal. The designer various medium such as multiple graphs, perceptions, lines, colours and textures, to explore more expression of modern lives with a conscience that is we return to life today.

设计师相信设计能带来"转变"的力量，设计可以将日常的平淡环境转变成特殊的空间体验，带领人们从某一种心情转变到另一种心情。设计的目的正是引领人们踏上一段充满"转变"的旅途，而这个目的是通过光影的运用、大小规模的组合和不同材质的交错达成的。设计师通过丰富的图形、比例、线条、色彩、质感等介质，以回归本心的坚韧，试图探索更多属于当代生活方式的表达——从今天起，我们回归生活。

THE FOCUS OF LIGHT LUXURY
轻奢聚焦

寻找手工艺的设计温度

设计师回避了这个时代的衍生物——工业化、标准化的产品，而选择了独特的、充满艺术性的家具，以期探索更多自我的表达，例如 La Cividina 的几何切割扶手椅、Alexander Lamout 荷花艺术造型蜡烛架、手工研制的高柜，还有主卧那灌注了无数个时代的匠人之心，凝聚着砂铸工艺的章鱼形把手。

融贯中西寻找雅致格调

6m 高的挑空客厅，以极致的当代设计美学呈现优雅的设计格调。地面渐变色的地毯，在丰富空间的同时不破坏空间的整体性；在强调现代感与艺术感的同时，也更多地保留舒适感。客厅家具在塑造包围感的同时，更是承载着一种多元文化的创意、革新与融合。

"江水漾西风，江花脱晚红。"朝南的江景餐厅与客厅、西厨及中厨相连通，在享受美食的同时，窗外的奢华尽收眼底。精致而克制的线条感铜制吊灯，金属曲线艺术装置与陈设艺术品，打破空间界限，使空间富有层次和现代感。

位于中心的深木旋转楼梯，如同艺术装置般立于其中，流畅的弧线呼应着建筑结构，传递着安静俊朗的空间气质。书房构建了一个既可独处也能容纳交流的空间。

减少需求，提高要求

主卧现代优雅的空间设计、素雅的基调、金色自然点缀、让每一个物件富有艺术灵性。当匠人的手划过雕刻的表面，堆砌着皱纹的双手有一种温柔，他的时间和心意都被按进了器物当中，为充满原始生命力的高柜注入了一丝细腻与温情。来自不同国家和品牌的家具以及艺术品，于这里巧妙邂逅，并在空间平衡出了自由的书写方式。

其他 3 间卧室，演绎不同的风格，创造新的感官体验。巧妙处理细节，将艺术贯穿其中，致力于追求独特审美，探索生活真谛。

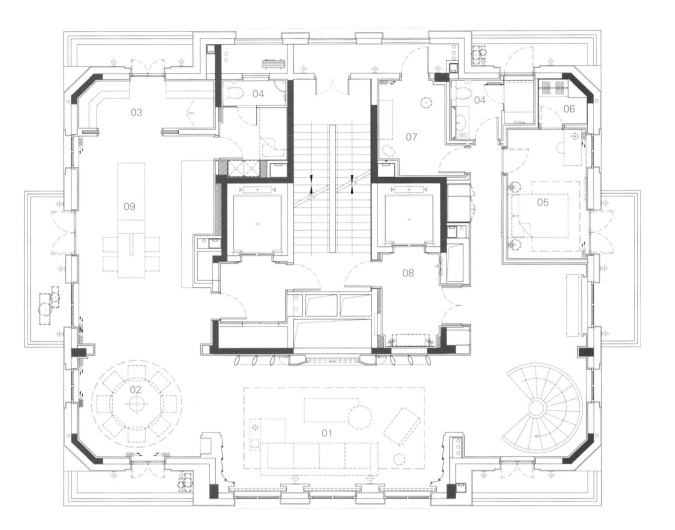

一层平面图
1st floor plan

01 客厅 / Living Room
02 餐厅 / Dining Room
03 厨房 / Kitchen
04 卫生间 / Bathroom
05 卧室 A / Bedroom A
06 衣帽间 / Walk-in Closet
07 拳击房 / Boxing Room
08 电梯厅 / Elevator Lobby
09 早餐厅 / Breakfast Area

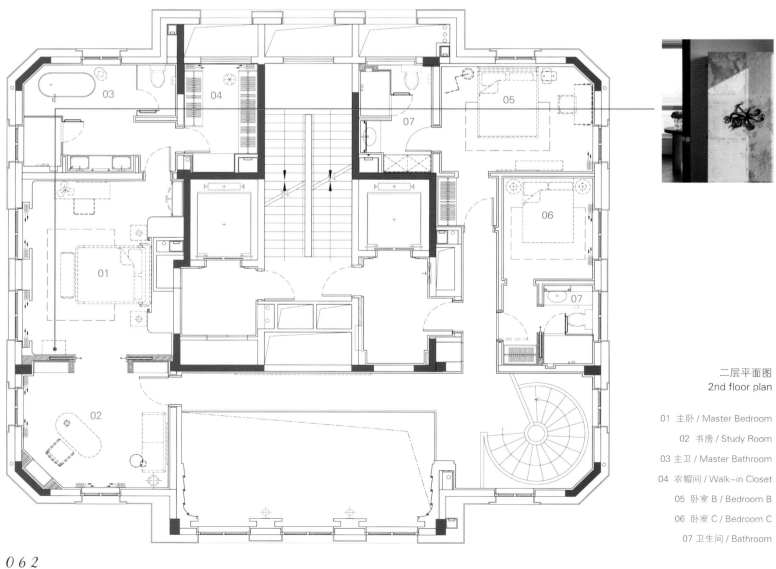

二层平面图
2nd floor plan

01 主卧 / Master Bedroom
02 书房 / Study Room
03 主卫 / Master Bathroom
04 衣帽间 / Walk-in Closet
05 卧室 B / Bedroom B
06 卧室 C / Bedroom C
07 卫生间 / Bathroom

灰色卧室大胆选用写意的艺术手绘墙纸，搭配玻璃、金属与皮质等现代材质家具，自然和谐，保持着栖息之所的感官平衡。

THE LIFE OF LIGHT LUXURY
轻奢生活

当你步入这里，想象自己是这公寓的主人时，仿佛体验着一种截然不同的生活。一定见过她，她无需依傍，只身在陌生的城市书写自己的旅程，那是杂糅着喜悦、伤痛与自由的旅程。周游世界并享受精致生活的人潮中，有她；在纽约购物、游览美术馆，与你擦身而过的人潮中，有她；在意大利科莫湖行进的游艇、在美丽的酒店里享用午餐的人潮中，有她。

ART MOVEMENT FLOWING IN WHITE SPACE

流动在白色空间里的艺术乐章

大陆轻奢 LIGHT LUXURY STYLE IN MAINLAND

● 项目信息　PROJECT INFORMATION

项目名称 / 白麓
设计公司 / STUDIO.Y 余颢凌设计事务所
主创设计 / 余颢凌
软装设计 / 刘芊妤
执行设计 / 阴倩

软装助理 / 唐竞
项目地点 / 四川成都
项目面积 / 1000 m²
摄影师 / 张骑麟

扫码查看电子版

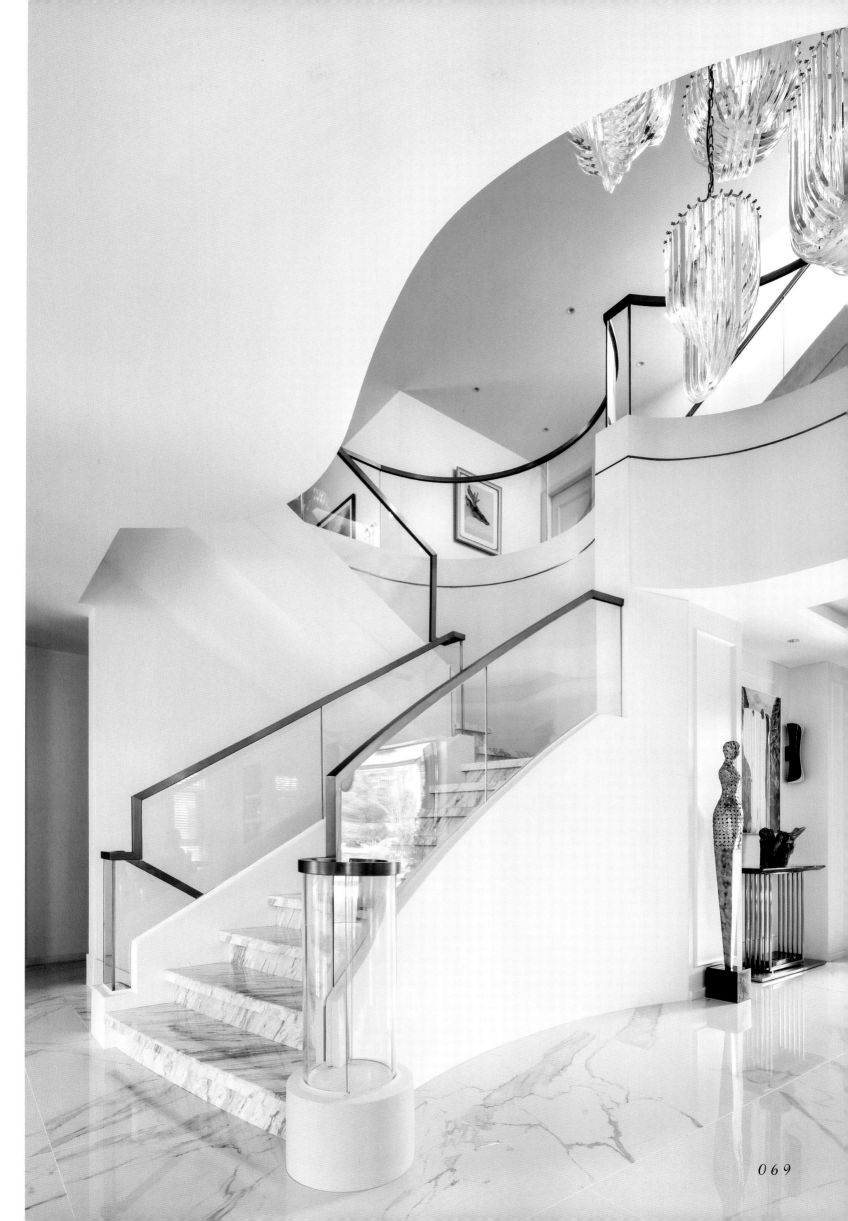

THE CONCEPT OF LIGHT LUXURY
轻奢理念

Kenya Hara, the writer of the book *White*, says, "The world is a luxury feast for all colours within your imagination. The fresh trees, the waves, the bright of fruits, the twinkle of the campfire, every colour are kind to all of us... White is the most distinctive and clear-cut image from the centre of chaos, and the colour white is an extreme example for peculiarity. It is not an entity from the combination, and it is not even a colour!" His sentences indicate the importance of white colour to aesthetic. The designer uses the white as the primary tone to represent clear and transparency for the entity. The designer uses this way to design the modern villa with the realism which is not the leading design styles for space design for traditional American villa and narrate a fair-sounding article.

原研哉在《白》中说："世界就像一场想象得到的种种颜色的奢华盛宴。树木的新鲜、水面的波光、水果的明丽、熊熊篝火的闪亮，这些颜色中的任何一种对我们来说都是亲切的……白是来自混沌中心最独特与鲜明的形象，白是这一独特性最极端的例子。它不是一种混合的实体，它甚至根本就不是一种颜色。"由此足见白色对美学营造而言是十分重要的存在。本案设计师以整体明净通透的白为主调，将以传统美式别墅为主且不符国人现代生活方式的空间设计为现代主义别墅，铺陈出一曲流动的艺术乐章。

1 美人鱼雕塑
2 视线一角，吊灯、摆件与花艺浑然一体，富有艺术感、精致感。

THE FOCUS OF LIGHT LUXURY
轻奢聚焦

在空间中融入生活，使之流动起来

入门即见的美人鱼雕塑造型优雅知性，白色蜿蜒而上的楼梯摒弃原有楼梯的尖角，修整弧度与线条的弯直，只留通透的玻璃与哑光金扶手。婉转回旋间，其仿佛永恒凝固的美好旋律，与作为入户与室内之间过渡带的钢琴厅谱出动静相谐的圆舞曲。

客厅延续白净的空间调性，以白色激发人们对于包容力、现代感、高级感等各种各样的联想。同时，设计师以天然石皮为局部墙体，使精良立体的质地彰显考究的硬朗气场。静卧的北极熊不能掩盖其体量与特质所带来的野性与刚烈本性，由此设计师将空间的场域与特定的生活方式架构于一体，营建差异化的独特空间气质。

日常的优美意识与形式感

模糊传统意义上的餐厅概念，根据特定的生活方式，设计两端为主人位并可同时容纳14人就餐的宴会厅，满足不定期的不同主题酒会派对需求，营造高级的生活仪式感。

知性与感性兼具

单纯的色块组合成温暖、闲适的主卧室，女儿房所用家居几乎全是温润的圆弧触面，墙面干净一色，家饰用品明丽清爽，整面的透明玻璃窗恰好引入无限的自然景致，让粉与白在日光的晕染下更加婉转可人。

为休闲、交谈、娱乐而打造的家庭厅是可以容纳多元存在的扩大化公共区，成为家庭成员情感互动的重要场所，完全中性的色彩氛围与家居造型也激发人找寻生活本来的意义。

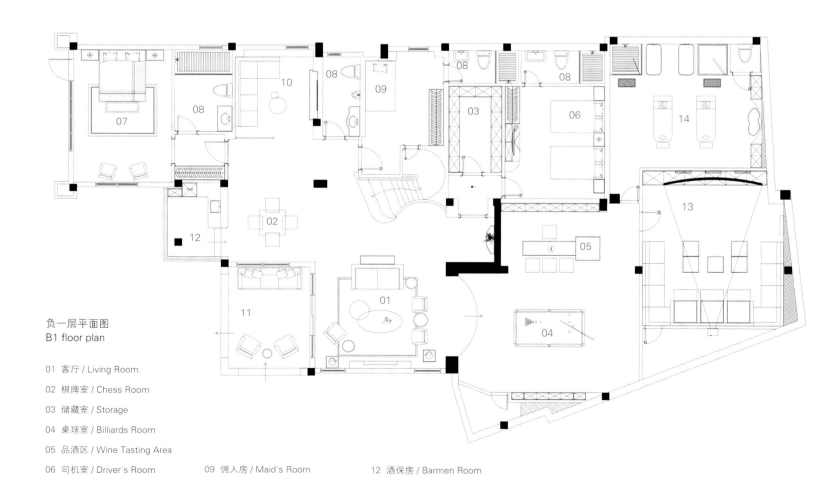

负一层平面图
B1 floor plan

01 客厅 / Living Room
02 棋牌室 / Chess Room
03 储藏室 / Storage
04 桌球室 / Billiards Room
05 品酒区 / Wine Tasting Area
06 司机室 / Driver's Room
07 客卧 / Guest Room
08 卫生间 / Bathroom
09 佣人房 / Maid's Room
10 司机休息室 / Driver's Lounge
11 露台 / Balcony
12 酒保房 / Barmen Room
13 家庭影院 / Home Theater
14 水疗室 / Spa Room

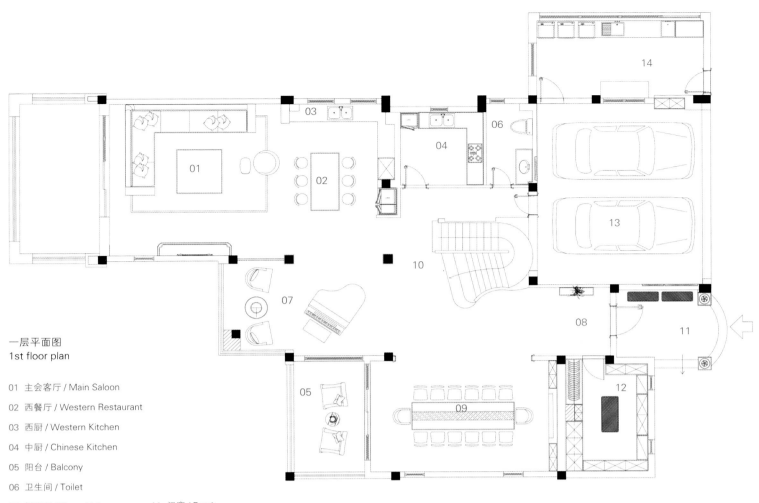

一层平面图
1st floor plan

01 主会客厅 / Main Saloon
02 西餐厅 / Western Restaurant
03 西厨 / Western Kitchen
04 中厨 / Chinese Kitchen
05 阳台 / Balcony
06 卫生间 / Toilet
07 钢琴厅 / Piano Hall
08 过道 / Corridor
09 宴会厅 / Banquet Hall
10 楼梯 / Staircase
11 门廊 / Porch
12 储藏室 & 衣橱 / Storage & Closet
13 车库 / Garage
14 洗衣房 / Laundry

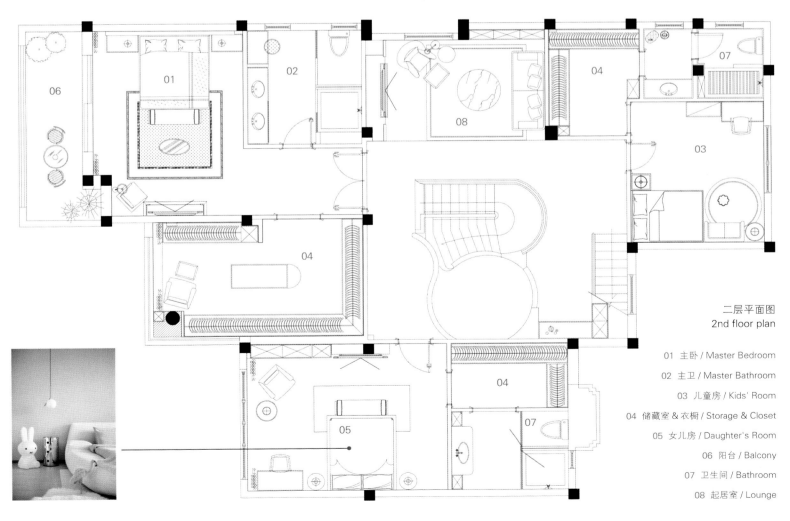

二层平面图
2nd floor plan

01 主卧 / Master Bedroom
02 主卫 / Master Bathroom
03 儿童房 / Kids' Room
04 储藏室 & 衣橱 / Storage & Closet
05 女儿房 / Daughter's Room
06 阳台 / Balcony
07 卫生间 / Bathroom
08 起居室 / Lounge

077

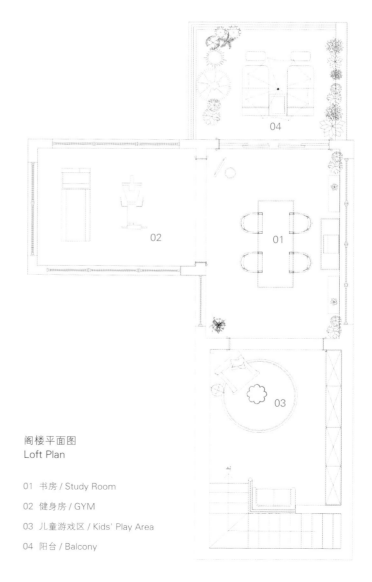

阁楼平面图
Loft Plan

01 书房 / Study Room

02 健身房 / GYM

03 儿童游戏区 / Kids' Play Area

04 阳台 / Balcony

THE LIFE OF
LIGHT LUXURY
轻奢生活

简单生活的要义，大多数时候在于居住者对空间的体验和互动关系，愈简单则愈有生命力。设计师运用柔软和中性的材质表现出极简，也在简约中传递细致的轻奢品位，唤起人们关于生活的点滴记忆，印刻着彼此的一颦一笑、一思一会，而生活的气息也在简单和轻奢的空间中变得饱满，充满意义。

THE ARTIST PLACE FOR VARIETY SHOW DIRECTOR

综艺导演的艺术栖所

LIGHT LUXURY STYLE IN MAINLAND

大陆轻奢

● 项目信息　PROJECT INFORMATION

项目名称 / 陆公馆
设计公司 / One House Design 壹舍设计
主创设计师 / 方磊
参与设计 / 金伟琳
视觉陈列 / 李文婷、李美萱、陈嫚
项目地点 / 上海
项目面积 / 190 m²
主要材料 / 石材、金属板、刻花玻璃、马来漆、皮革等
摄影师 / 隋思聪

扫码查看电子版

083

客厅、餐厅立面图
Living Room, Dining Room

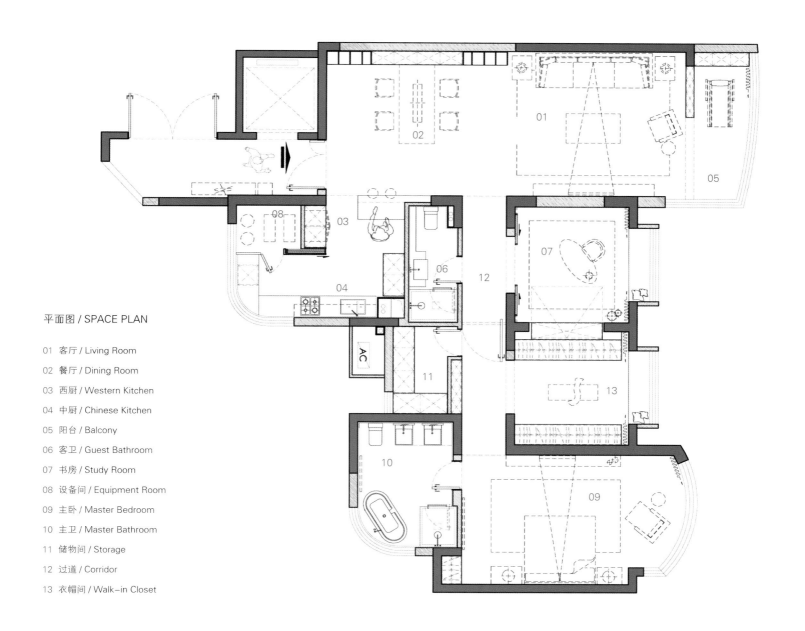

平面图 / SPACE PLAN

- 01 客厅 / Living Room
- 02 餐厅 / Dining Room
- 03 西厨 / Western Kitchen
- 04 中厨 / Chinese Kitchen
- 05 阳台 / Balcony
- 06 客卫 / Guest Bathroom
- 07 书房 / Study Room
- 08 设备间 / Equipment Room
- 09 主卧 / Master Bedroom
- 10 主卫 / Master Bathroom
- 11 储物间 / Storage
- 12 过道 / Corridor
- 13 衣帽间 / Walk-in Closet

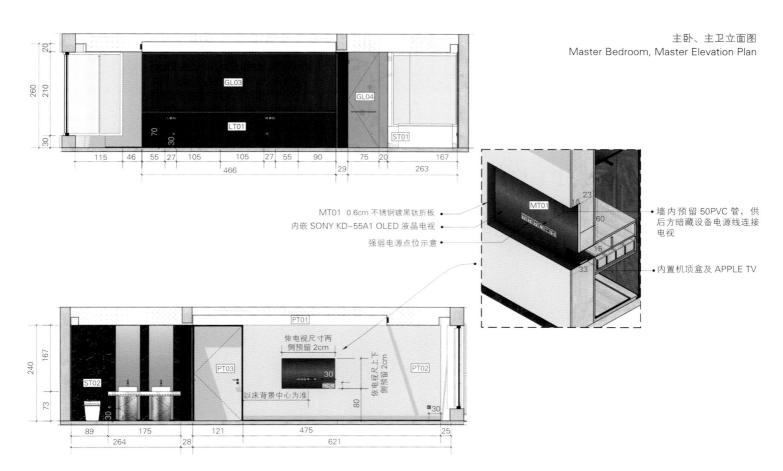

主卧、主卫立面图
Master Bedroom, Master Elevation Plan

- MT01 0.6cm 不锈钢镀黑钛折板
- 内嵌 SONY KD-55A1 OLED 液晶电视
- 强弱电源点位示意
- 墙内预留 50PVC 管，供后方暗藏设备电源线连接电视
- 内置机顶盒及 APPLE TV

THE CONCEPT OF
LIGHT LUXURY
轻奢理念

Lens records the sublimation of language, which is far more vivid than the text. Lu Wei, the host of this house and the vice President of Canxing production, who has successfully created many popular programs such as street dance, the voice of China, the colourful Chinese and the masked singers. The director hides the picture he wants to capture in every frame which seems like an understatement, but it can make the emotion more vividly. While the designer integrates the artistic expression of the lens into space and polishes every corner with a gentle and practised attitude, the visual beauty and the style connotation coexist which is in accord with the expectation of the residence.

With the zero thinking when writing a script, which explores the maximum possibility of space and develops a sense of surprise at the right time. In the structural transformation, the pursuit of sunlight, air and scenery can flow freely and inject a sense of hierarchy and nature. Simple decoration, in a calm and restrained tone, creates space fusion and visual tension. The designer outlines his design themes in this way.

镜头记录是语言的升华，远比文字来得鲜活。居者是灿星制作副总裁陆伟，成功打造《这！就是街舞》《中国好声音》《出彩中国人》《蒙面唱将》等火爆节目。导演将想要捕捉的画面藏匿于每一帧，看似轻描淡写，却能让感情呼之欲出，而设计师将镜头的艺术表达融入空间，以从容洗练的态度精心打磨空间每一隅，视觉美感与格调内涵交织并存，与居者的期待一拍即合。

"以剧本创作时的归零思维，发掘空间最大可能性，并适时衍生惊喜感。在结构改造中，追求阳光、空气、风景能自由畅快流动，注入层次感和自然感。摆设从简，在平静且克制的基调中，创造空间融合及视觉张力。"设计师如此概述其设计主旨。

THE FOCUS OF LIGHT LUXURY

轻奢聚焦

从玄关进入便是开放式格局，客餐厅与厨房一览无余。饱含珍重记忆与情意的长桌是居者心爱之物，与鱼骨拼地板相得益彰，旧物件和谐融入新空间，营造出别致韵味。

沙发线条利落，塑造大气沉稳风范。运用落地金属格架的虚实相间划分出阳台，既呈现轻盈穿透的视觉效果，又让客厅面宽和布置配比更合理。绿植和阳光相遇，将自然气息注入其中。

嵌入墙壁的定制格架，最大化利用面宽有限的书房空间，辅以内置灯带，氤氲出现代美学气息，满足居者阅读或创作诉求。步入式衣帽间，雅致的烟灰色玻璃柜门，不仅开阔空间视野，也方便日常搭配检视。

主卧位于走廊尽头，床头立面的刻花玻璃和客厅电视背景墙遥相呼应，一侧灯饰、家俬采用设计师方磊惯用的非对称三角构成手法，演绎多种材质的碰撞、光与影的错落交叉。而窗外的绿植，也成为室内最美好的点缀。

浴室偌大的弧形拐角玻璃窗，模糊室内外界限，带来充足的采光，愈发突显洁净清透。边沐浴边欣赏庭院的景色，压力和疲倦即刻荡然无存。

1 抽象性艺术挂画
2 轻巧独特的床头灯
3 卧室：黑白灰的经典配色，以金属配件点缀空间

←

遍布于餐厅、客厅、书房以及衣帽间对面的格架，相映成趣，方便居者随手取阅书籍或置物。"根据居者特定的生活方式量身设计，因势运用每一寸空间，拿捏每一处细节，让一切都显得刚刚好。"设计师补充。

THE LIFE OF LIGHT LUXURY

轻奢生活

再是忙碌的工作，终有平静的时刻；再是奔波的生活，总要有落脚的居所。在苍鹰飞累了之后，栖息的地方不过求一片安宁舒适。

设计师坦言："不同于导演在节目中所表现出的喧闹热烈，这个家的气质展现是如此安宁内敛的，张弛之间，映射出居者的生活情趣以及视野格局。这无疑能给予他更多舒适和放松，让他乐享惬意时光。"

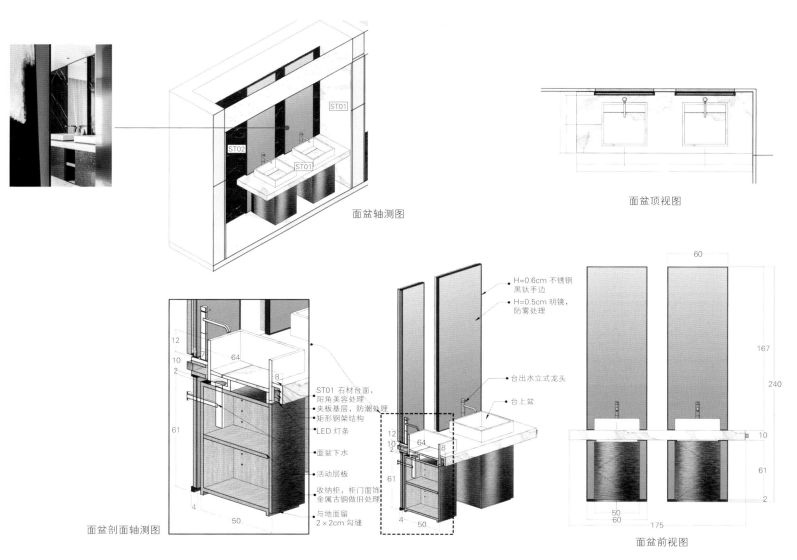

面盆轴测图

面盆顶视图

- H=0.6cm 不锈钢黑钛手边
- H=0.5cm 明镜，防雾处理

台出水立式龙头

台上盆

ST01 石材台面，阳角美容处理

夹板基层，防潮处理

矩形钢架结构

LED 灯条

面盆下水

活动层板

收纳柜，柜门面饰金属古铜做旧处理

与地面留 2×2cm 勾缝

面盆剖面轴测图

面盆前视图

SOAKING INTO LIGHT LUXURY AND CULTIVATING ELEGANCE

浸润于轻奢，陶冶出优雅

LIGHT LUXURY STYLE IN MAINLAND
大陆轻奢

● 项目信息　PROJECT INFORMATION

项目名称 / 深圳京基·御景中央样板房
设计公司 / H DESIGN
设计师 / 胡坤
设计团队 / 姜洋、吴常朴、张巍、郭媚
软装陈设 / TT Design
项目地点 / 广东深圳
项目面积 / 97 m²
主要材料 / 皮革、爵士白大理石、黑钢、深胡桃木等
摄影师 / B+M Studio. 彦铭

扫码查看电子版

THE CONCEPT OF LIGHT LUXURY

轻奢理念

Geneviève Antoine-Dariaux, a leading authority of France fashion, wrote in her book A Guide to Elegance that people should live in an elegant and noble life through all kinds of living areas. She says, " Elegant is a harmony, a sense of beauty and the gift from the god, while elegant is the production of art. Elegant comes from the cultural moulding, and it also develops from the cultural moulding."

In these case, the designer uses the concept of "Modern Delicate Live" and annotates some new phrases such as "Luxury", "Fashion" and "style". The designer pays attention to delicate details, the collocation of colour, high quality of selecting the suitable material and whether this design has both senses of classic and modern or not. The designer uses dark grey marble, solid wood combined twills and teak shelf to break the lonesome which makes the spacial level is vividly portrayed and gives the audience a sense of elegance.

法国时尚界泰斗安东丽·德阿里奥夫人在《优雅》一书中详尽地描绘了如何从生活的方方面面去塑造优雅高贵的生活方式，"优雅是一种和谐，类似于美丽，只不过美丽是上天的恩赐，而优雅是艺术的产物；优雅从文化的陶冶中产生，也在文化的陶冶中发展"。

本案中，设计师以"现代精致生活"为追求理念，对"豪华""流行""时尚"一类的词汇进行了新的诠释，在乎细节的精致、色彩的搭配、选材是否高品质、设计是否兼备经典与现代感，以深灰色纹路大理石、实木复合斜纹地板、柚木色搁架打破寂寥，令空间的层次感呼之欲出，上演极致上品的优雅生活。

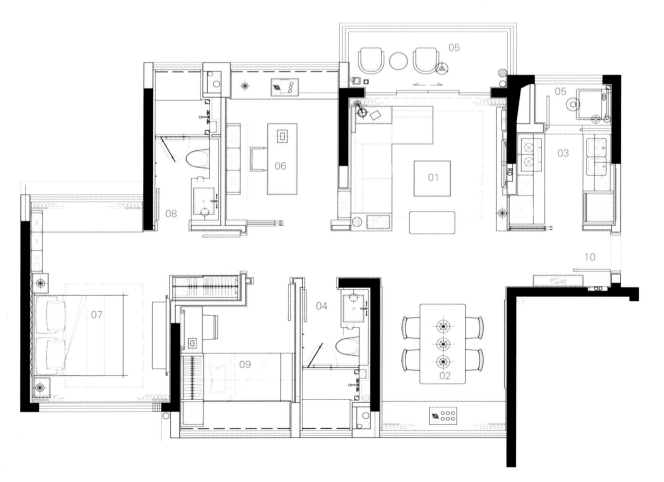

平面图 / SPACE PLAN

01 客厅 / Living Room
02 餐厅 / Dining Room
03 厨房 / Kitchen
04 卫生间 / Bathroom
05 阳台 / Balcony
06 书房 / Study Room
07 主卧 / Master Bedroom
08 主卫 / Master Bathroom
09 儿童房 / Kid's Room
10 玄关 / Foyer

THE FOCUS OF LIGHT LUXURY
轻奢聚焦

通过入户门厅步入家庭生活主要区域，客厅与餐厅的设计异曲同工，同样的中轴对称、同样的奢华却不浮夸的尺度。由内到外，流畅的线条在空间里相互碰撞、交汇，释放出一种静谧和惬意，促成"归家团聚"的美好。

材料的质感叙述了空间的关系，极富触感的织物、大理石台面、皮质座面，每一个精彩的细节都可以成就非凡设计。在设计师看来，设计需要不断地突破与思考，培养对设计的认知和对概念的坚持，一个好的作品在满足使用者需求的同时，也要找到其中的平衡点，兼顾其艺术性与创造性。

南北通透的空间布局让人在白天中的任何时候都能享受到日光的沐浴，同时室内不再强调主光源的需求，设计师以射灯和艺术灯作为陪衬，塑造利落、精致的空间风骨。

1 艺术灯：选用的吊灯简约、大气，线条利落，增添餐厅的时尚氛围

2 柚木色搁架：柚木色契合空间主打的气质与质感，使得搁架既有实用的置物收纳功能，也有美化作用。

THE LIFE OF
LIGHT LUXURY
轻奢生活

优雅，人们首先想到的可能是可可·香奈儿或是克里斯汀·迪奥所创造的各种漂亮、优雅形象，她们着装精致，气质知性，举止大方而尽显高贵。在空间中，注入现代人居生活的精致与优雅，以深沉色泽的家具为主，构成与干净而雅致的界面相得益彰的效果。当阳光透过玻璃纱窗洒进室内，很自然地生成了一派岁月静好的景象，优雅的气场与轻奢的态度兼顾，这便是对灵魂深处所求的家的细心反馈。

THE SIMPLICITY AND ELEGANT SPIRIT OF THE EXTREME PENTHOUSE

极致顶复的简雅精神

大陆轻奢 LIGHT LUXURY STYLE IN MAINLAND

● 项目信息　PROJECT INFORMATION

项目名称 / 融创 北京壹号院　　　参与设计 / 周晟、邵俊兵、张卫
设计公司 / DIA 丹健国际　　　　项目地点 / 北京
空间设计 / 张健　　　　　　　　主要材料 / 雅典娜灰、珊瑚海、染色木皮、不锈钢等
陈设艺术 / 谈翼鹏　　　　　　　摄影师 / 罗文

扫码查看电子版

The designer abandons many ways to design traditional villa such as elaborated decoration and splendid and magnificent, the designer adopts the simplicity and elegant style to highlight the sense of elite. He has a keen interest in architecture, so he uses many architecture languages in his design to emphasise the concept of "Inside Architecture" which is the critical value of Danjian International Group that is to break the decoration design from its surface and focus on the idea of space design.

设计师摈弃了传统豪宅设计惯用的装饰繁复、金碧辉煌等手法，采用国际化的简雅风格，突显空间的精英品位。设计师对建筑有强烈兴趣，在室内大空间设计中善于运用建筑语言，以实现"内建筑"为设计——这一突破装饰表皮设计，关注空间设计的理念，是丹健国际的重要核心价值。

THE CONCEPT OF LIGHT LUXURY 轻奢理念

THE FOCUS OF LIGHT LUXURY
轻奢聚焦

建筑设计在玻璃幕墙式住宅和大平层官邸的成熟理念基础上进行了创新，利用曲线穹顶的设计搭建出顶层的复式空间。设计师在复式上层室内的设计中充分利用这一独有的建筑特色。

设计中大量使用亚光面、皮革、茶色金属以及布艺软装，配合素色材质，融入国际化的设计手法，强调精英品位的同时，着力打造空间的舒适性与实用性，使豪宅内"家"的概念真正回归到了对舒适的现代化生活的追求中。

玄关、客厅、餐厅这三个功能区围绕室外庭院依次展开，黑色与白色体块在庭院室内围墙处产生交错穿插关系，强调了特色庭院在整个空间的主导地位，体现出设计师对空间的高度敏感性。层叠的天花及简雅高冷的配色强化了穹顶的存在，最大化引入室外怡人景色。

复式下层主要为私人领域的卧室、卫生间、起居室，设计手法以体现舒适性、私密性为主。材质的选择多用染色木皮、地毯等，配色为给人宁静感的灰色及棕色，带来舒适安静的感受。

1 厨房的西餐吧台
2 复式下层楼梯间
3 现代简约家具

THE LIFE OF LIGHT LUXURY
轻奢生活

　　财富的积累与社会地位的提升给了许多人足够的自信，人们不再需要张扬的、繁琐的、炫耀式的符号来突显自己，反而更加关注自身精神世界的充实。由金碧辉煌到时尚高端，由表象富足到内心富足，由纸醉金迷的对外张扬到体贴入微的人本关怀，私密、温暖、现代感、国际化以及独特个性成为当代人对居所的核心要求。

↑ 复式上层平面图　upper floor plan

01　客厅 / Living Room　　　07　玄关 / Foyer
02　餐厅 / Dining Room　　　08　电梯厅 / Elevator Lobby
03　厨房 / Kitchen　　　　　09　楼梯 / Staircase
04　卫生间 / Bathroom　　　 10　佣人房 / Maid's Room
05　阳台 / Balcony　　　　　11　西餐吧台 / Western Kitchen Bar
06　书房 / Study Room　　　 12　设备平台 / Equipment Flat

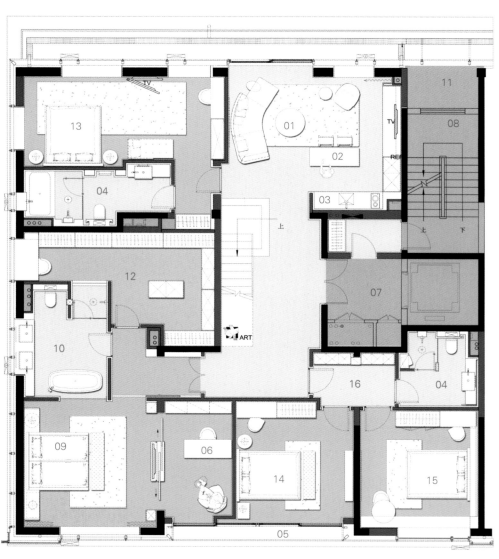

← 复式下层平面图　lower floor plan

01　起居室 / Setting Room　　09　主卧 / Master Bedroom
02　早餐台 / Breakfast Table　10　主卫 / Master Bathroom
03　西厨 / Western Kitchen　 11　设备平台 / Equipment Flat
04　卫生间 / Bathroom　　　　12　衣帽间 / Walk-in Closet
05　阳台 / Balcony　　　　　 13　父母房 / Parents' Room
06　书房 / Study Room　　　　14　卧室 A / Bedroom A
07　电梯厅 / Elevator Lobby　 15　卧室 B / Bedroom B
08　楼梯 / Staircase　　　　　16　子女套房 / Children's Suite

ENJOYING THE ELEGANT FOUR SEASONS · WINTER

坐拥四季的清雅·冬

大陆轻奢 LIGHT LUXURY STYLE IN MAINLAND

● 项目信息　PROJECT INFORMATION

项目名称 / 万科·安亭湖语森林别墅样板房　　**设计团队** / 刘雅娟、廖蕾
设计公司 / MAUDE 上海牧笛设计　　　　　　**项目地点** / 上海
设计总监 / 毛明镜　　　　　　　　　　　　**项目面积** / 200 m²

扫码查看电子版

THE CONCEPT OF LIGHT LUXURY
轻奢理念

The housing estate to which this villa belonged has a natural lake, and this natural lake looks like a rare blue diamond which has a sense of elegant, peace and steady. Based on this advantage, the designer creates a living place with birds' twitter and fragrance of flowers and is full of vigour. The theme of this design is the "Winter", and the designer uses warm white, grey and aqua blue which represents the peace and elegance of the lake as its main tones.

别墅所在小区内拥有一池天然湖，犹如一颗稀世罕见的蓝宝石，散发着高雅、安静、沉稳的气息。设计师借此自然优势，打造出一处生机勃勃、鸟语花香的四季景象之生活居所。设计出发点是以四季之"冬"为主题，主体色调取自冬季的雪、被赋予阳光的暖白、大地的灰，以及那湖中高雅、宁静的水蓝色。

 # THE FOCUS OF LIGHT LUXURY
轻奢聚焦

空间设计手法以纵向面延展，利用线性收边创造面与线的完美结合，赋予了简单空间以多层次变化。整体空间配以考究的家具、精致的饰品以及合理的摆放，呈现一个为注重生活品质和格调的精英人士所定制的居家环境。

置身于客、餐厅空间，5.5m的挑高空间带来的开阔与震撼感扑面而来。挑高的背景墙整体采用白色天然大理石，利用其独有的天然纹路，以精湛的工艺对纹手法，打造出一幅宛如江南烟雨缥缈的朦胧画卷，与现代感十足的壁炉完美结合，舞动的火苗能温暖整个冬季。

紧邻餐厅区域转身便是开放式厨房，将厨房与餐厅、客厅连接起来，不但可以使居家空间更加通透宽敞，还增加了互动性。厨房配有吧台，坐在吧台旁边喝着酒，那就非常惬意了。

二层、三层是卧室空间。在3层的主卧置入独立卫生间、书房及强大的衣帽收纳系统，充分满足居者对生活的高品质要求。

→ 现代简约壁炉

THE LIFE OF
LIGHT LUXURY
轻奢生活

在厨房吧台处，享受下午茶的惬意或沉醉在鸡尾酒的微醺中，配以舒适温暖的场景，实现人们对未来美好生活的遐想。新时代的生活，是富足而内涵的，低调而充实的，不论财富多寡、地位高低，都能在生活中找到自己的闪光点。

→ 抽象性摆件

LIGHT LUXURY HONGKONG

○ 现代轻奢 / Modern Light Luxury

STYLE IN

港式轻奢
LIGHT LUXURY STYLE IN HONGKONG

轻奢理念 / The Concept of Light Luxury
轻奢聚焦 / The Focus of Light Luxury
轻奢生活 / The Life of Light Luxury

轻奢印象
THE IMPRESSION OF LIGHT LUXURY

"香港学"研究学者洪清田先生曾表示，香港是中国把"中体西用"做得最成功的城市。独特的地理环境、历史文化、政治经济都深刻影响着香港的室内设计，造就了独具风味的"港式设计"。港式轻奢在原始港式设计基础上删繁就简，保留传统港式设计"商务精英"的空间气氛，引入国际潮流，创造个性化港式新时尚。它将古典与现代、传统与时尚的元素兼容并蓄，多元化元素在空间中既对立又统一，是一种透露着对高品质生活的理解与尊重的设计格调。

Hong Qingtian, a scholar of "HK studies" said, HK is the successful city which applies to "Westernised Chinese Style". HK interior design is influenced by the unique geographical conditions, history, culture, politics and economy, which makes the unique "HK Design Style". Light luxury style in HongKong simplifies from the original design, and keeps traditional feature for "Business Elite". It also introduces international trends to create an Individualised HK new fashion. It combines the classical and modern, traditional and fashionable elements with each other; and the diversity of elements are opposite and unified in space, which shows a design style of understanding and respect for high-quality life.

1 郑炳坤｜澳门私人住宅
开放式厨房、客厅

→ 1

01 轻奢理念 THE CONCEPT OF LIGHT LUXURY

港式轻奢以后现代主义风格为基础环境，室内设计具有历史的延续性，但不拘泥于传统的思维方法，设计章法中流露出温馨的人情味，试图创造一种融感性与理性、集传统与现代、糅大众化与专业性于一体的独特空间。

HK light luxury is based on Post-modernism, and the interior design has the historical continuity, but not limited the traditional thinking. The design reveals a warm human interest and trying to create a unique space that combines sensibility and rationality, tradition and modernity, publicity and professionalism.

02 轻奢空间 THE SPACE OF LIGHT LUXURY

空间设计不仅注重居室的实用性，更注重在商业化社会生活中的精致时尚与个性化。利用多种不同的材质组合空间，常以色彩、陈设来实现不同区域的特定功能和轻奢韵味，明亮或暗淡、华丽或古朴、平滑或粗糙，相互穿插对比。多元化的设计手法灵活多变，空间简洁自然，形成有力量但不生硬，有活力但不稚嫩的家居形象。

The space design emphasises not only the practicability of the living room, but also the exquisite fashion and personalisation in the commercial social life. Using a variety of different materials combinations, and utilising colours and furnishings to reach specific functions and affordable luxury in different areas, they are interspersed with bright or dull, gorgeous or quaint, smooth or rough each other. The Diversified design is flexible, space is concise and natural, which forms a home image with strength but not stiff, energetic but not tender.

03 轻奢色彩 THE COLOUR OF LIGHT LUXURY

配色理性、线条简单，常见古铜色、孔雀蓝、香槟金等色调，并运用大量的中间色系让中西元素互相融合，以金属质感、晶莹水晶等装饰来突出港式风味的轻奢气质。

Proper colour matching and simple lines, such as common bronze, peacock blue, champagne gold and other colours which it can use a large number of neutral colours to blend the Chinese and Western elements so that it highlights the affordable luxury of HK style temperament under decorations of metal texture, bright crystal.

04 轻奢用材 THE MATERIAL OF LIGHT LUXURY

常见材料有抛光石材、瓷砖、暖色木材、玻璃等，配饰材质常见水晶、皮革、镜面以及金属工艺品等。

The common materials of the affordable luxury are polished stone, ceramic tiles, warm wood and glass. The accessories contain of the affordable luxury are crystal, leather, mirror and metal crafts.

05 家具配饰 FURNITURE ACCESSORIES

通常采用线条简洁流畅、造型时尚的包豪斯风格家具，繁简结合，重视家具的触感和家具搭配的视觉效果。

It adopts stylish Bauhaus style furniture with simple and smooth lines. It combines complexity with simplicity, paying attention to the touch of furniture and the visual effect of furniture matching.

2 郑炳坤｜元朗独立屋
  一层客厅

3 尚策室内设计顾问有限公司｜上海金巢铂瑞阁公寓 A 户型
  餐厅

4 香港泛纳设计事务所｜水合院
  客厅

NOT TALKING ABOUT LUXURY, WE ONLY TALK ABOUT LIFE

不言奢华，只言生活

港式轻奢 LIGHT LUXURY STYLE IN HONGKONG

● 项目信息　PROJECT INFORMATION

项目名称 / 水合院　　　　　　　项目地点 / 广东深圳
设计公司 / 香港泛纳设计事务所　　项目面积 / 1500 m²
设计团队 / 潘鸿彬、莫定轩、王晓川　摄影师 / 吴潇峰

扫码查看电子版

THE CONCEPT OF LIGHT LUXURY

轻奢理念

This 1500m² fifth-floor single-family villa model room is located in Shenzhen Guangdong, the spatial planning refers to the traditional structure of quadrangle dwellings, and the designer designs this villa based on this structure. Each floor has its unique function, and all of them are planned around the central yard. Upward, people can see the sky and downstairs, and people can see a swimming pool. Though it is indoors, people still can see the sky, the water, trees, moon and stars which brings the residence an enjoyable experience that they are living in nature. The designer knows how to design this vast space with moderate luxury, refuses to pile up mediocrity and excess luxury in that villa and chooses a significant amount of wood to neutralise the sense of luxury. The modest luxury does not be a minor issue taking precedence over a major one, and it gives owners more freedom to enjoy their lives.

这是一座位于广东深圳的别墅样板房，五层楼的独立别墅占地1500m²，空间规划参照传统中国四合院格局，以此重新演绎。每层楼具有不同的功能区，围绕中央的"庭院"进行空间规划。上有天幕，下有泳池，虽在室内亦有天空、绿水、林木和星月等美景，给住户如生活在大自然中的奇妙体验。对于阔绰大气的空间，设计师懂得把握奢华的分寸，拒绝平庸与过度奢华的堆砌，选择大量木色与木材中和奢华感。于是，适度的轻奢没有了喧宾夺主的气势，更多的是留给屋主享受生活的自由。

一层平面图
1st floor plan

01 客厅 / Living Room
02 餐厅 / Dining Room
03 厨房 / Kitchen
04 卫生间 / Bathroom
05 室内泳池 / Indoor Swimming Pool
06 户外按摩水池 / Outdoor Massage Pool
07 休闲区 / Leisure Area
08 品茶区 / Tea Area
09 庭院 / Courtyard
10 艺术空间 / Art Space

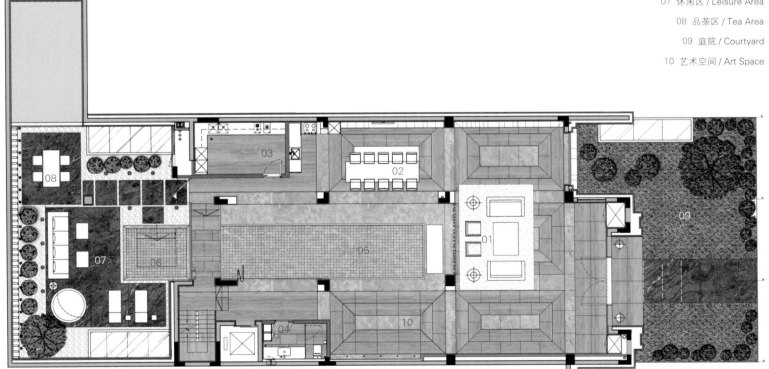

 # THE FOCUS OF LIGHT LUXURY
轻奢聚焦

首层客厅和餐厅围绕着中央运动区的室内泳池来布局。两层高的客厅和池边的用餐区呈开放式与厨房相连，机动天窗将室外的光线引入整栋楼，新鲜空气亦能畅通流入。

在夏日，房子中央的池水能调节室内温度而节约能源，晚间还能将星光与月色倒映在水面。客房和主人房设于一楼和二楼两边，以垂直的落地木帘包围，提高私隐度。开放式平面布局设计了滑动趟门，使主人房室内外的分界变得模糊，以木材和白灰色大理石为主，配以暖色调家具，使空间显得柔和及摩登。

休闲和娱乐功能区位于负一、二层，具备更佳的隔音效果。中庭泳池区有两层楼高，中空上方挂有铜管 LED 灯，为派对空间呈现出星光灿烂的宴会气氛，再度绘出了大自然的景致之美，宾主在浪漫的气息中进行阅读、品尝美酒、观看影剧、修身健体，度过欢乐的家庭时光。

→ 1

1 主卧套间：开放式的衣帽间和卫浴空间，宽敞而便捷，木色调更是增加空间温暖气息。

143

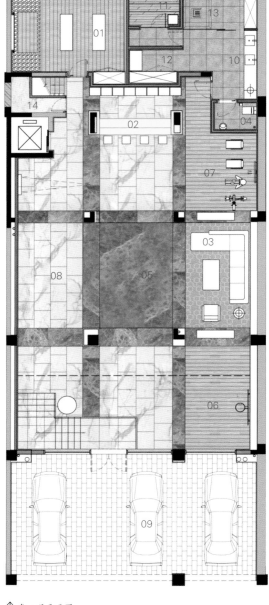

负一层平面图
B1 floor plan

01 影音室 / Video Room
02 品酒区 / Wine Tasting Area
03 阅读区 / Reading Area
04 卫生间 / Toilet
05 工人房 / Workers' Room
06 洗衣房 / Laundry
07 机房 / Engine Room
08 中空 / Hollow
09 工人卫生间 / Workers' Bathroom

负二层平面图
B2 floor plan

01 瑜伽房 / Yoga Room
02 水吧 / Water Bar
03 休息区 / Resting Area
04 卫生间 / Bathroom
05 多功能区 / Multifunction Area
06 运动区 / Sporting Area
07 健身房 / GYM
08 艺术空间 / Art Space
09 车库 / Garage
10 水疗区 / SPA Area
11 桑拿房 / Sauna Room
12 更衣区 / Dressing Room
13 淋浴间 / Shower Room
14 设备间 / Equipment Room

↑ 负一层平面图   ↑ 负二层平面图

THE LIFE OF LIGHT LUXURY
轻奢生活

五层独栋别墅所塑造的生活并不是土豪式的过度享受，而是师法天地，师法自然，以舒适为前提，打造富有弹性的生活空间。"月盈则亏，水满则溢"，任何过度追求的奢华装饰，到头来只会降低对美好生活的感知度，于是设计师将传统观念通过现代手法和语言发扬光大，给予现代奢华以新的定义。轻奢格调在满足人们对奢华追求的同时，又不失关注生活本质的初心，让一切在这里找到平衡感。

二层平面图
2nd floor plan

01 卧室A / Bedroom A
02 卧室B / Bedroom B
03 卧室C / Bedroom C
04 卫生间 / Bathroom
05 衣帽间 / Walk-in Closet
06 偏厅 / Side Hall
07 中空 / Hollow

三层平面图
3rd floor plan

01 主卧 / Master Bedroom
02 主卫 / Master Bathroom
03 阳台 / Balcony
04 户外休息区 / Outdoor Leisure Area
05 中空 / Hollow
06 衣帽间 / Walk-in Closet
07 偏厅 / Side Hall
08 美容区 / Beauty Area
09 卧室 D / Bedroom D
10 户外休息阳台 / Outdoor Leisure Balcony
11 造景区 / Landscape Area
12 半户外浴室 / Semi-outdoor Bathroom

LIGHT DESIGN, HIGH FASHION

轻质设计　高感时尚

LIGHT LUXURY STYLE IN HONGKONG

港式轻奢

● 项目信息　PROJECT INFORMATION

项目名称 / 澳门私人住宅
设计公司 / Danny Cheng Interiors Ltd.
设计师 / 郑炳坤

项目地点 / 澳门氹仔
项目面积 / 604 m²
主要材料 / 水泥、石、镜、木、金属元素等

扫码查看电子版

This high-end residence in Macao faces north and faces the sea with the hills for a background. In the beginning, the keyword of this design is Modern Minimalist. The designer uses black, white and grey tones in some light and transparent elements such as metal, reflector lamp and glasses to manifest elegant touch of Affordable Luxury, the different shapes of furniture and decorations to show the unique and characteristic of this high-end residence. The spacial design focuses on finding fashion energy in the clam and creates fashion-forward tension in the low profile.

坐南向北，背山面海，地处澳门的这一高端住宅在设计初始便围绕着关键词"现代简约"进行设计构思。黑、白、灰的时尚色调在金属、射灯、玻璃等轻质通透的现代化元素中彰显轻奢品质的雅致格调，造型各异的家具、饰品塑造了住宅的独特与个性。空间设计在沉稳中寻找时尚的活力，在低调中创造前卫的张力。

THE CONCEPT OF LIGHT LUXURY

轻奢理念

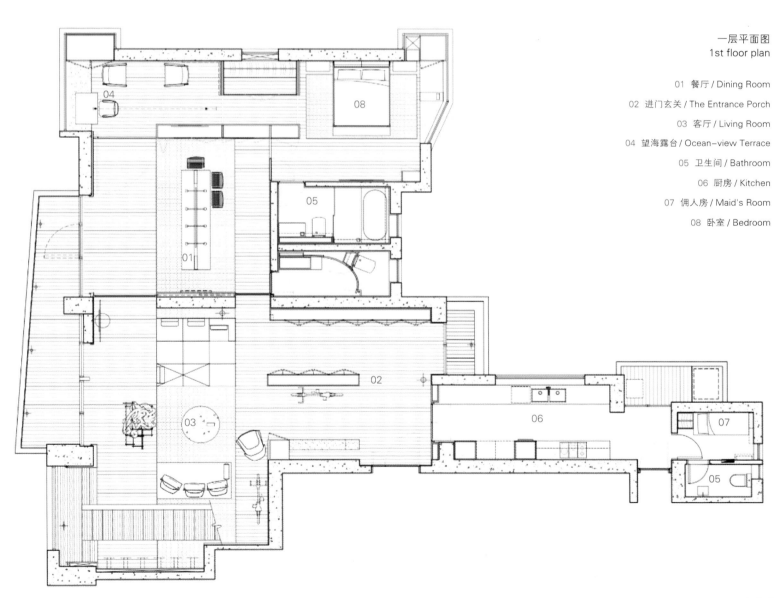

一层平面图
1st floor plan

01 餐厅 / Dining Room
02 进门玄关 / The Entrance Porch
03 客厅 / Living Room
04 望海露台 / Ocean-view Terrace
05 卫生间 / Bathroom
06 厨房 / Kitchen
07 佣人房 / Maid's Room
08 卧室 / Bedroom

室内运用大量玻璃、镜面元素，使得室内都市感十足，为置身其中的人营造了一种前卫的空间气氛。

玄关处设置镜面墙，打破空间的单调，并开阔了玄关的视觉感受，给人通透开朗的空间感。

二层走廊以玻璃作护栏隔断，不但维护了上下层之间的空间统一性，还保证了二层的采光，配合造型各异、错落有致的沙发组合更显住宅的摩登时尚感。

→ 1
金属质感的器材配置星空地毯,使得空间更具科幻现代感。

THE FOCUS OF LIGHT LUXURY
轻奢聚焦

简约设计，提升住宅整体空间感

一进门便能感受到高端住宅的奢享雅致，南北走向的设计让人踏进家门就能欣赏到极致海景；镜面玄关给人层次分明的视觉感，塑造了通透、利落的空间体验；两层楼高的天花板造就了开阔明朗的空间。

客厅的灰白墙面能在阳光照射下呈现出不同的光影变化，散发出现代设计的简洁魅力；银色水滴形藏品置于墙壁，串联空间的同时极大地丰富了空间的艺术性。

黑白色调，营造空间的摩登时尚感

餐厅深色实木背景墙塑造强烈的视觉冲击，镜材餐桌与极具张力的地毯相互映衬，使空间以震撼的整体性构成一种仿佛置身于宇宙的科幻的空间感。

以金属材质的阶梯连接楼层，嵌于墙上的黑色书架起伏之间创造出奇妙的动感，空间在现代与前卫之间记录着物质的美妙与精神的富足。

金属质感，打造轻奢华私人空间

二楼是居者的个人领域，开放式的主卧浴室设计大胆前卫，随性而自由。作为卧室背景的灰镜展示着其独特质感，极尽现代时尚的光泽度令主卧焕发轻度奢华的涵雅气质。浴室大面积铺陈大理石，石材纹理清晰，尽显自然美感，配合明暗有致的灯光，空间分外清爽大方。

1 金属质感的楼梯多了一丝现代酷感，镜面效果在折射光线的同时，也令空间更为宽敞。

2 主卧以黑白灰元素为主，搭配金属感装饰，呈现简约工业风。

THE LIFE OF LIGHT LUXURY
轻奢生活

艺术设计因冲突而丰富；生命长河因波澜而生动。下层餐厅旁由艺术家 Peter Anton 创作的"巧克力"艺术品是轻质设计中的视觉焦点，演绎着一个众所周知的秘密——"生活就像一盒巧克力，你永远不知道下一块是什么"。

↓ 战舰模型点缀走廊，呼应星空地毯的科幻想象。

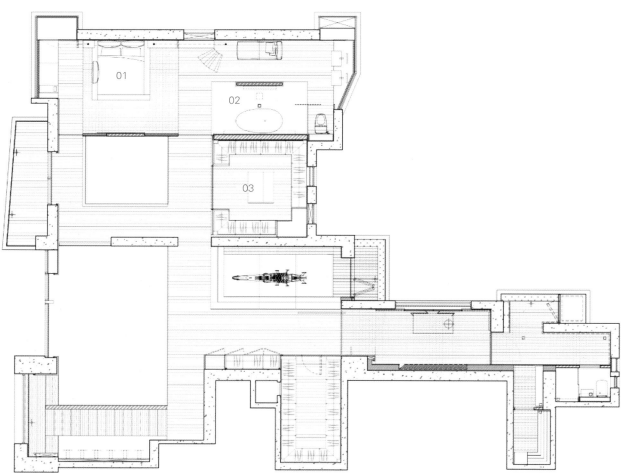

二层平面图
2nd floor plan

01 主卧 / Master Bedroom
02 主卫 / Master Bathroom
03 衣帽间 / Walk-in Closet

USING TRADITIONAL DESIGN TO BUILD MODERN HOUSE

以传统设计章法打造现代住宅

港式轻奢 LIGHT LUXURY STYLE IN HONGKONG

● **项目信息**　PROJECT INFORMATION

项目名称 / 云端总裁公馆

设计公司 / CCD 香港郑中设计事务所　　　**项目地点** / 湖北武汉

扫码查看电子版

The theme of this flat is based on the "Towering as mountain Tai and flowing as rivers" which comes from the classic Chinese Qin, and the designer uses the literary quotation of Yu Boya and Zhong Ziqi to reveal landscapes of mountains and rivers and the sense of spiritual communion.

The reception area of this design which continues the concept that is the designer uses many scenes of classic eastern yards to represent different functional areas such as the reception entrance, the rest area, the art space and the private spaces which allows people to feel like they are in the high mountain and flowing water.

公寓以中国古琴曲《高山流水》中的"峨峨兮若泰山，洋洋兮若江河"作为设计主题，借用"昔伯牙鼓琴，子期能知其曲中高山流水之意，两人遂结成知音"的典故，在空间中展现山环水绕、情意相通的融融之情境。

前厅接待区域的设计从项目整体的设计理念延续而来，即现代建筑外壳内的"亭台与回廊"：将接待入口、休息区、艺术空间、私人空间等多个功能区域设计成东方古典院落中的场景，好似悠远的居住情境融于高山流水之间。

THE CONCEPT OF LIGHT LUXURY
轻奢理念

　　设计师在入口处划分出独立的前厅接待空间，将通往不同套房的通道设计为亭台回廊，一路上以艺术品与装置小景观重现古代的竹林小径，引导宾客缓缓步入，如同中国古代若有客来访，主人定当引领他穿过三进院落下榻至厢房。

　　休息区座位疏密分布，有的布置在朝向落地窗景前，可直面长江天幕，阳光透过玻璃幕墙投射在大地色的地毯上，形成斑驳如树阴的疏影。

　　空间明暗、开阖的节奏也是项目的关键之一。在设计师看来，这种空间和氛围的变化是中国传统空间最具魅力的地方。因此，设计营造出来的空间序列在自然光和人工照明之间交替，空间的明暗也跟随自然光和人工照明的交替而变化；与之同步的是，视线的通透、封闭、半通透也在这个过程中被精心安排。人们进入这里，在不同时间、不同角度，会看到古典"院落围合"的重现，仿若感知环境与古曲《高山流水》的契合。

轻奢聚焦 THE FOCUS OF LIGHT LUXURY

THE LIFE OF LIGHT LUXURY

轻奢生活

子曰:"知者乐水,仁者乐山;知者动,仁者静;知者乐,仁者寿。"古曲《高山流水》蕴含天地之浩远,山水之灵韵,包含了"天人合一""物我两忘"的传统文化精神。取水之敏捷包容,山之沉着冷静,处高位而不自扰,居浮华而能自清——修身养性是毕生之学。

ON THE BUSTLING LIFE AND HAVING A BEAUTIFUL HOUSE

繁华之上，浮沉随心

港式轻奢 LIGHT LUXURY STYLE IN HONGKONG

● 项目信息　PROJECT INFORMATION

项目名称 / 上海金巢铂瑞阁公寓 A 户型
设计公司 / 尚策室内设计顾问有限公司
主案设计师 / 李奇恩、康谭珍
参与设计师 / 吴婉芝

项目地点 / 上海
项目面积 / 134 m²
主要材料 / 雅士白、银貂灰、卡拉白玉石、灰橡木饰面、灰橡木地板等
摄影师 / 陈思

扫码查看电子版

THE CONCEPT OF LIGHT LUXURY

轻奢理念

The Shanghai Jinchaobo Ruigebin Apartment locates at the busy area of Dapuqiao which is one of the commercial districts of Shanghai. The Penthouse Flat has the best geographic location to look far and overlook Magic City. Because of locating in a high place, the designer wants the owner to enjoy and fully take advantage of their house with his designs. So the primary tones of this design are black, white and grey, the simplicity and not losing grace tone cannot only emphasise the function of this design but also show the steady atmosphere for this design.

上海金巢铂瑞阁公寓坐落于打浦桥繁华地段，是上海中心商业街区之一。顶层复式公寓，拥有绝佳的地理位置远眺、俯瞰魔都。身居高处，设计师考虑的是如何让设计突出高层住宅带给业主居住上的优势与享受。于是整体空间色调以黑、白、灰为主，简约而不失风雅的色调不仅仅能强调功能性，更是尽显高处空间的稳重大气。

THE FOCUS OF LIGHT LUXURY
轻奢聚焦

以精致的现代轻奢风格作为定位，为这座高处公寓设计增添更多亮点。客厅电视背景墙和水吧台均采用了卡拉白玉石，通透的石材、天然的纹理与前卫的风格浑然一体尽显主人品位；餐厅餐台与户外餐台相互连通，从室内延伸到室外，为体验时尚都市增添了新亮点；有凹凸感的墙纸，为棋牌室增添了层次感；而金属马赛克墙身，直立圆柱洗手盆的主人卫生间，更是令人眼前一亮。

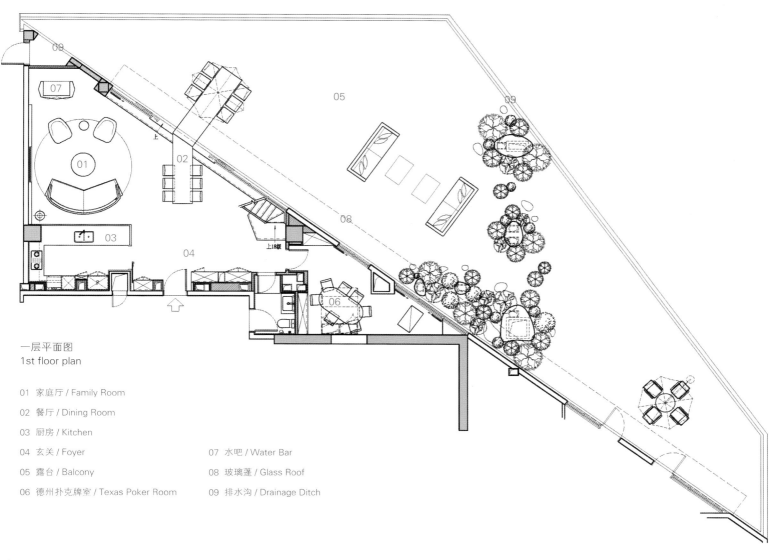

一层平面图
1st floor plan

01 家庭厅 / Family Room
02 餐厅 / Dining Room
03 厨房 / Kitchen
04 玄关 / Foyer
05 露台 / Balcony
06 德州扑克牌室 / Texas Poker Room
07 水吧 / Water Bar
08 玻璃蓬 / Glass Roof
09 排水沟 / Drainage Ditch

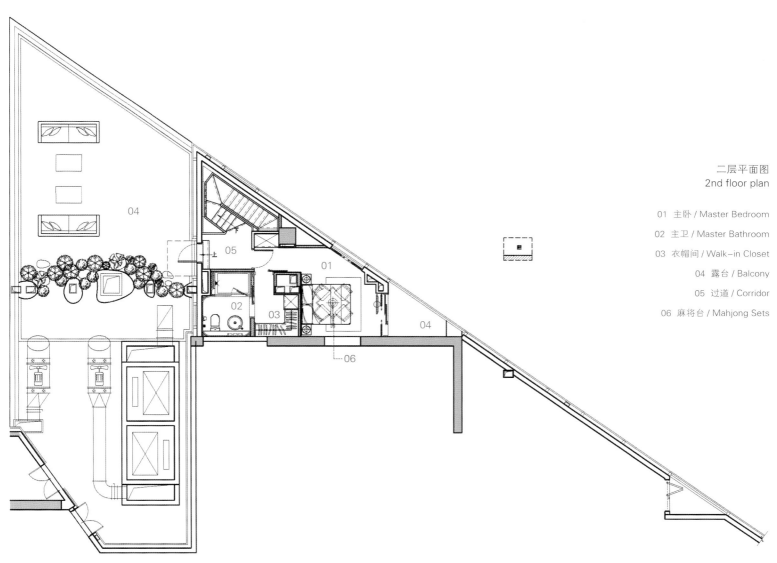

二层平面图
2nd floor plan

01 主卧 / Master Bedroom
02 主卫 / Master Bathroom
03 衣帽间 / Walk-in Closet
04 露台 / Balcony
05 过道 / Corridor
06 麻将台 / Mahjong Sets

THE LIFE OF
LIGHT LUXURY
轻奢生活

生活在繁杂多变的都市里已是烦扰不休，而简单的生活空间却能让人身心舒畅。顶层的复式公寓身处在城市的繁华之中，别树一帜的轻奢设计，既能拥有享受高品质生活的格调，又凭借着楼高优势暂能与城市喧嚣保持一定距离，让人沉淀下来感受到宁静与舒适，让躁动的心得以安抚平静。

DELICACY, SIMPLICITY AND LUXURY IN HEART

精致，
简奢于心

LIGHT LUXURY STYLE IN HONGKONG
港式轻奢

● 项目信息　PROJECT INFORMATION

项目名称 / 元朗独立屋
设计公司 / Danny Cheng Interiors Ltd.
设计师 / 郑炳坤
项目地点 / 香港元朗

项目面积 / 372 m²
主要材料 / 云石、墙纸、铝板、木纹地板、户外木、地砖等
摄影师 / Bobby Wu

扫码查看电子版

This is the Danny Cheng's house, and he uses the concept of "Simple is beautiful" incisively and vividly. In this space design, the designer chooses the colour white as its main tone and uses grey tone as the support element to create the spacial level which allows this place more spacious and crystal clear. Also, the designer uses many neutral tints such as black, grey and brown to mix and decorate which would enable the pure white does not look so pure, though it is the primary colour, it does not mean it is rigid.

这是设计师郑炳坤的家，在这里，设计师将其"简单即是美"的设计理念演绎得淋漓尽致。空间设计以白色为主，灰色辅助，创造空间层次，使宽敞之余多了一份清透。此外，辅之以黑、灰、褐等中性色调进行调和与点缀，令纯白不苍白，单色不呆板。

THE CONCEPT OF LIGHT LUXURY
轻奢理念

一层平面图
1st floor plan

01 客厅 / Living Room
02 餐厅 / Dining Room
03 厨房 / Kitchen
04 阳台 / Balcony
05 卫生间 / Bathroom
06 卧室 / Bedroom
07 储藏室 / Storage
08 工作区 / Working Area
09 佣人房 / Maid's Room
10 (高尔夫) 轻击区 / Putting Room

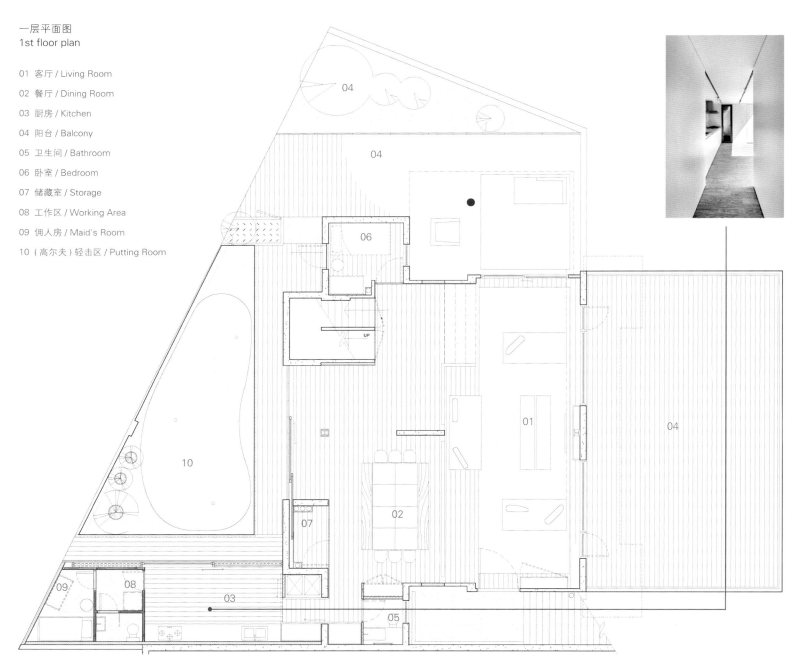

负一层平面图
B1 floor plan

01 客厅 / Living Room
02 车库 / Garage
03 储藏室 / Storage
04 阳台 / Balcony
05 卫生间 / Bathroom
06 练习场 / Driving Range
07 休息厅 / Lounge Hall
08 按摩房 / Massage Room

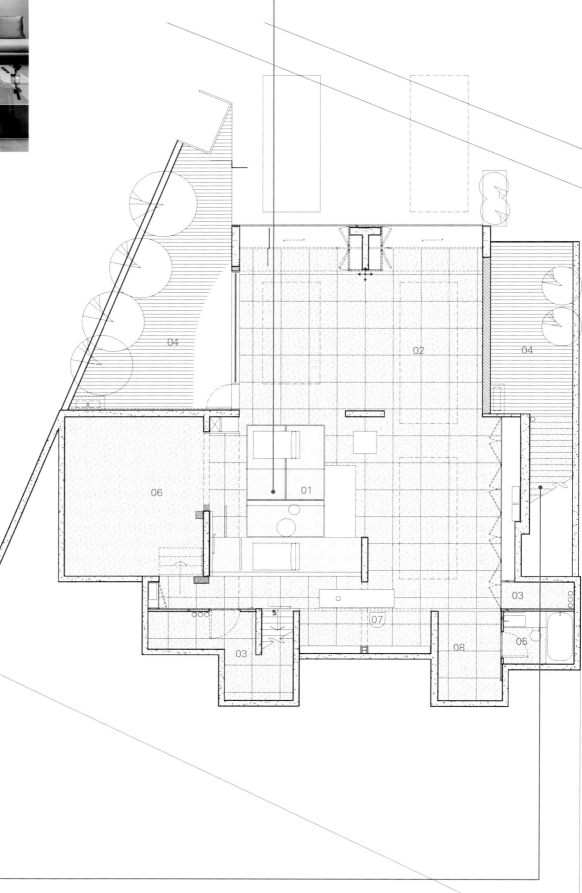

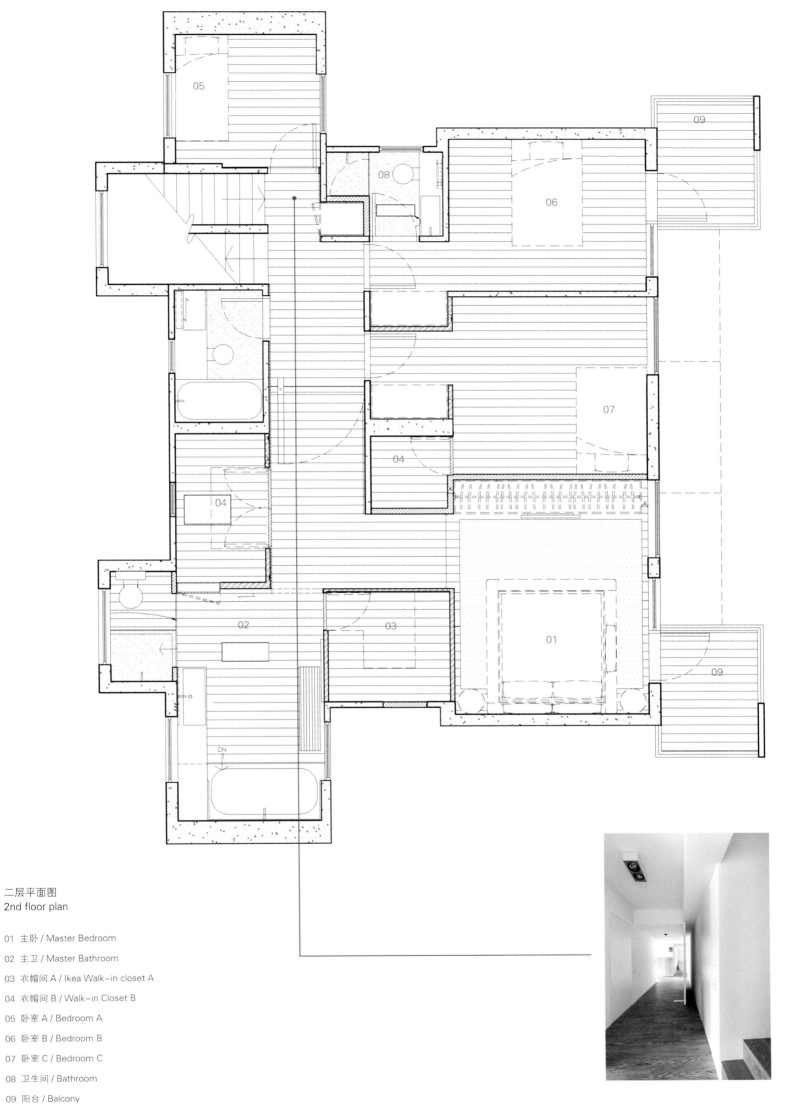

二层平面图
2nd floor plan

01 主卧 / Master Bedroom
02 主卫 / Master Bathroom
03 衣帽间 A / Ikea Walk-in closet A
04 衣帽间 B / Walk-in Closet B
05 卧室 A / Bedroom A
06 卧室 B / Bedroom B
07 卧室 C / Bedroom C
08 卫生间 / Bathroom
09 阳台 / Balcony

1 大自然本是最好的设计,莫兰迪色系配色的家居就足以勾勒它的辽阔景致。

THE FOCUS OF LIGHT LUXURY
轻奢聚焦

白色为灵魂,串联了空间

客厅中,浅色木地板与细纹外墙交相呼应,桌椅、装饰则都是饱和度较低的浅色,砖面、抱枕则是深灰色,另有黑色元素点缀其中,不仅使得白色之家有了生机与活力,更增加了空间的层次感和纵深感。最富巧思的是一组沙发,除了同为冷色调与地面、墙面相呼应,不规则造型柔和了建筑线条带来的刚硬与冰冷。

简单即是美,前卫现代风

打破旧常规,车库与客厅仅玻璃之隔,车库成了精致的展览厅,赋予其观赏性与实用性。游泳池烘托自然景致,它不仅勾连了室内与室外,更与大面积普照的阳光融为一体。

室内整体风格突显现代感,而室外更加亲近大自然。大型绿植是后院的亮点,姿态蓬勃的褐色枝干是后院最好的装饰,与线条平直的浅色桌椅相映成趣;枝头绽放的点点绿意,则为房子增添了不同于室内的精彩与生气。

 # THE LIFE OF LIGHT LUXURY
轻奢生活 ─────○

寻一处栖息之所安放所有的情绪与情感，简约的空间设计给每一颗心预留足够的容量。简洁线条勾勒宁静的生活，精致家具拼凑亲密的感情，呼朋唤友相聚于此，打球游泳、品酒赏车，轻奢生活是能够品味人情的美好与精神的丰满。

THE FADED GLITZ RETURN TO THE NATURE, AND FEELING THE COZY AND SIMPLE LIFE

褪去浮华 舒享于简

LIGHT LUXURY STYLE IN HONGKONG

港式轻奢

● 项目信息　PROJECT INFORMATION

设计公司 / Millimeter Interior Design Limited
设计师 / Michael Liu　　**项目面积** / 435 m²
项目地点 / 香港新界　　**主要材料** / 水泥、木板、墙纸、木地板、喷漆、角铁等

扫码查看电子版

THE CONCEPT OF LIGHT LUXURY

轻奢理念

This is a 435m² house located in the suburb of HongKong, and it is a relaxed, quiet and fashion house. The designer uses modern white walls and wooden floor which is warm and attractive. The transparent design for the garage made it become the dream house and allow people to enjoy the beautiful scenes even in the living room. The designer changes the stairs to artistic steel stairs not only reduce the visual interference but also allow people to see the beautiful garden and at the same time, increase the natural light indoors.

这座面积约为435m²的房子坐落在香港郊区，是一座轻盈、宁静而又时尚的住宅。设计师采用现代白色墙壁与木质地板，温暖而又有吸引力。车库的透明设计使它成为热情车主的梦想之家，他们可以随时在客厅欣赏他的收藏。设计师将楼梯换成了美观的铁质楼梯，不仅减少了对爱车的视觉干扰，同时还纳入庭院的景致，增加室内自然光线。

负一层平面图
B1 floor plan

一层平面图
1st floor plan

二层平面图
2nd floor plan

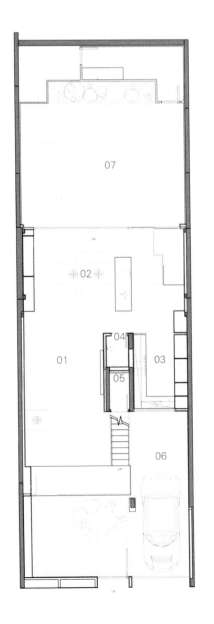
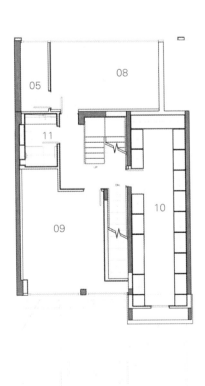
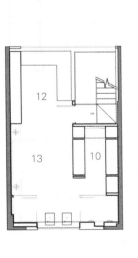

01 客厅 / Living Room
02 开放式厨房 / Open Kitchen
03 厨房 / Kitchen
04 客卫 / Guest Toilet
05 储藏室 / Storage
06 车库 / Garage
07 花园 / Garden
08 卧室 A / Bedroom A
09 卧室 B / Bedroom B
10 衣帽间 / Walk-in Closet
11 儿童浴室 / Kid's Bathroom
12 书房 / Study Room
13 主卧 / Master Bedroom

THE FOCUS OF LIGHT LUXURY
轻奢聚焦

室内设计简洁，家具饰品较少，很好地保持了空间的洁净感，适度的留白也给予屋主更多自由的空间。黑白灰的布艺沙发与带着柔和曲线的茶几，形成有趣的空间串联。客餐两厅以玻璃区分隔离庭院空间，庭院景观可融入室内而成为重点景观之一，增添悠然从容的生活气息。餐厨区一黑一白的餐桌与中岛台，互为对比。完全开放的空间以简洁的设计，让生活的品质感更上一层楼。

二楼作为主人房楼层，功能性完备，包含卧室、书房、衣帽间及浴室。白色调为主加入灰色布艺，凝聚柔和温暖的氛围。书桌前偌大的落地玻璃窗，增加空间采光与视野界线，一览无余的外景，也能让人更加神清气爽地工作。

THE LIFE OF LIGHT LUXURY
轻奢生活

位于郊区的住宅，能让人尽情享受属于自己的安静生活。屋主注重室内采光和窗外风景的互动融合，一方庭院将生活的脚步延伸至户外。屋主有喜欢的跑车和收藏品，与家人隐于郊外，室内足够的空间大大提升了生活品质感，而简约的轻奢便是都市人理想生活的最好模样，不失格调又舒适自在。

SIMPLE AND EXQUISITE DESIGN, WHICH MAKING FAINT JOY AND TRANQUIL LIFE

简致设计合围人生清欢

LIGHT LUXURY STYLE IN HONGKONG
港式轻奢

● 项目信息　PROJECT INFORMATION

项目名称 / 加多利园
设计公司 / Danny Cheng Interiors Ltd.
设计师 / 郑炳坤

项目地点 / 香港元朗
项目面积 / 300 m²
主要材料 / 云石、墙纸、人造白橡木板、防水泥板等

扫码查看电子版

THE CONCEPT OF LIGHT LUXURY

轻奢理念

This is a 300m² villa located in Yuanlang, Hong Kong. Mies van der Rohe, the master of modernism in architecture in the middle of the 20th Century, has a well-known saying, "Less is more." It can be described as the soul of the modernism in architecture. This designer uses this aesthetic thought which annotates the true essence of minimalism and creates an affordable luxury life which satisfies the appreciation of the beauty of modern people.

该案例是香港元朗一处面积约为300m²的豪宅。二十世纪中期最著名的现代主义建筑大师密斯·凡德罗的一句名言："少即是多（Less is more）"，可以说是现代设计的核心灵魂。设计师将这一美学思想糅入家居设计中，诠释着极简的真谛和打造更符合现代人审美的轻奢品质生活。

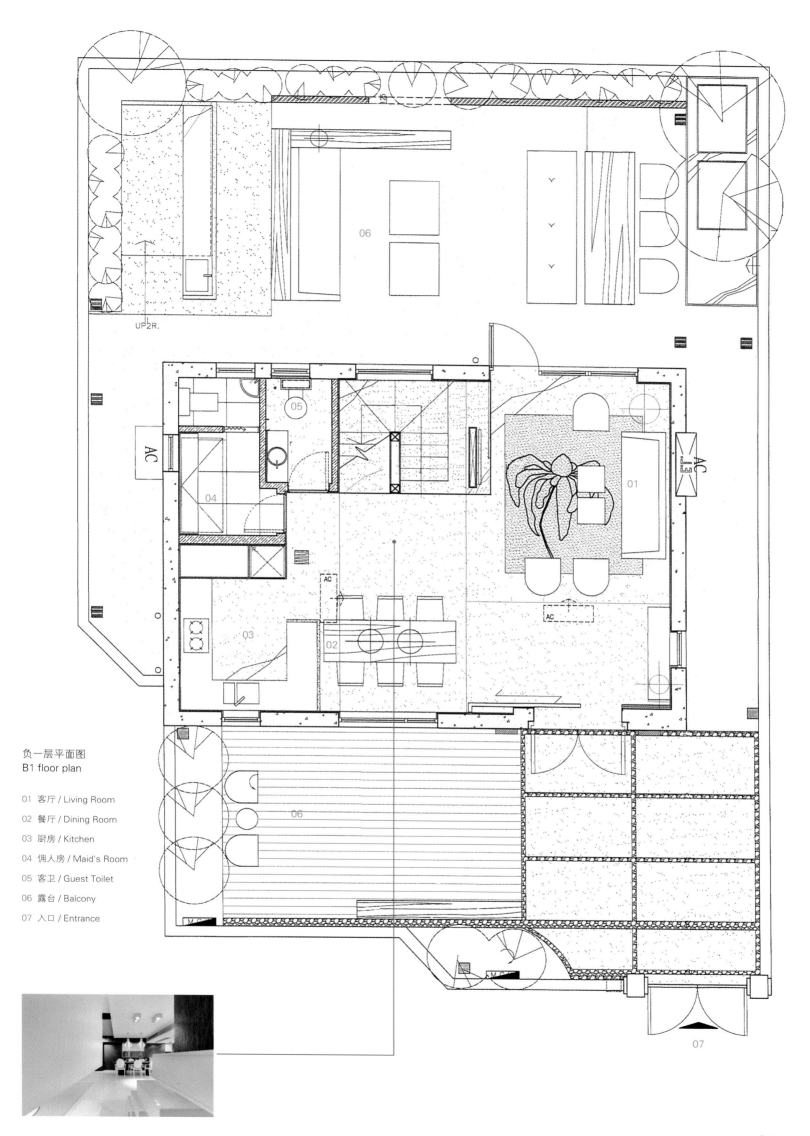

负一层平面图
B1 floor plan

01 客厅 / Living Room
02 餐厅 / Dining Room
03 厨房 / Kitchen
04 佣人房 / Maid's Room
05 客卫 / Guest Toilet
06 露台 / Balcony
07 入口 / Entrance

THE FOCUS OF LIGHT LUXURY
轻奢聚焦

曲直线条，对比强烈

在一个空间中，隔断和摆设越多，对生活的束缚也就越大，而这套家居设计化繁为简，余留下的元素将空间彰显得明亮通透。餐厅厨房白色台面与洁白的大理石地板交相呼应，朴实浑厚的深棕色木质墙壁一扫极简空间的冷酷。木色与灰白色的搭配，看似淡雅的色彩与结构，却在不经意间为住户带来温暖的舒适之感。客厅的用色比例与平衡布局，让极简充满层次又富有活力，巧妙地放置简洁的纯白色软装进行搭配，不仅为空间增添了灵动之感，同时也是一种独特的气质体现。

去冗除杂，删繁就简

一进大门，白灰的天花板配合明亮的自然光线创造出通透开阔的空间感。首先映入眼帘的是极具线条感的鱼骨型旋转楼梯，线条干净利落。配合四角翘起的弧形沙发宛如展现女人的曼妙身姿，两者强烈的对比碰撞出了强烈的艺术美感，令人沉浸其中。

巧妙引入，光影变幻

设计师通过对灯光的处理，巧妙地分割出各个功能区域。偌大的落地玻璃窗肆意地让自然光在空间中穿梭，伴着白昼的更替，散发出不同角度不同强度的光晕，丰富了住户的入住体验。庭院绿意盎然的植物点缀，造出郁郁葱葱的户外景观。步入主卧，玻璃落地窗外探出的丝丝绿意，打破了空间的沉闷感，让住户拥抱来自清晨的第一缕阳光。而书房是一隅净化心灵的居心地，窗内白色搭配原木色，干净雅致。窗外草色入帘青，是住户静心潜读的不二之选。

1 现代楼梯：极具线条感的楼梯，镂空设计更显其干净利落的造型。

2 楼梯和廊道连接着开放式的书房，更是一种空间上无需多言的交流与对话。

轻奢生活
THE LIFE OF LIGHT LUXURY

没有冗繁的装饰，一切从简，正是设计师打造的轻奢极简主义风格。谁说豪宅必须用繁冗来堆砌，轻度的奢华舒适与极简亦能诠释生活本真。正所谓"雪沫乳花浮午盏，蓼茸蒿笋试春盘，人间有味是清欢"。

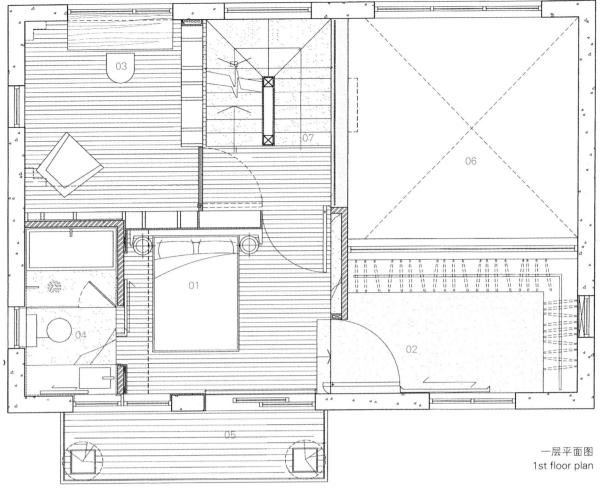

一层平面图
1st floor plan

01 卧室 / Bedroom
02 衣帽间 / Walk-in Closet
03 书房 / Study Room
04 卫生间 / Bathroom
05 阳台 / Balcony
06 中空 / Hollow
07 楼梯 / Staircase

二层平面图
2nd floor plan

01 主卧 / Master Bedroom
02 主卫 / Master Bathroom
03 衣帽间 / Walk-in Closet
04 卧室 / Bedroom
05 卫生间 / Bathroom

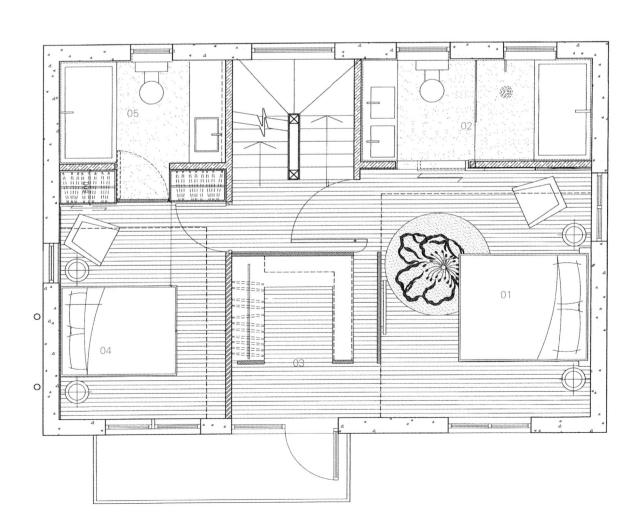

顶层平面图
Top Floor Plan

01 露天花园 / Rool Garden
02 露天沙发 / Outdoor Sofa
03 按摩浴池 / Jacuzzi
04 水盘 / Water Pond
05 人造木地板 / Artificial Wooden Floor
06 咖啡桌 / Coffee Table

LIVING IN A LAND OF IDYLLIC BEAUTY

身心栖居的『世外桃源』

LIGHT LUXURY STYLE IN HONGKONG

港式轻奢

● 项目信息　PROJECT INFORMATION

项目名称 / 名镌别墅
设计公司 / 香港泛纳设计事务所
主创设计师 / 潘鸿彬
设计团队 / 郑珮华、黄颖彤
项目地点 / 广东深圳
项目面积 / 1000 m²
摄影师 / 吴潇峰

扫码查看电子版

This private villa is located in Shenzhen, and the design aims to enhance the harmony between the indoor space and natural environment within the original concept of this villa. The ground floor has a living room and a dining room, and the basement level has many entertainment rooms such as a yoga room, a wine cellar and an outdoor swimming pool. The top floor is the master room and the guest room. The main elements of this whole villa are the visual relation of vertical and horizontal and the penetration of natural lights so that it can bedim the boundary of the space and create a suitable integration living space.

此私人别墅项目坐落于深圳，设计手法强调在原本建筑理念中增进室内空间与自然环境的融入与和谐。地面层包括起居室及餐厅，负一层有瑜伽房、酒窖及室外泳池等娱乐休闲空间，顶层为主人房与客房。贯穿整栋别墅的主要元素是垂直与水平的视觉联系和自然光线的穿透，目的是使所有空间的界线模糊化，以打造更为一体化的舒适居住空间。

轻奢理念
THE CONCEPT OF LIGHT LUXURY

负一层平面图
B1 floor plan

01 客厅 / Living Room
02 露天泳池 / Outdoor Swimming Pool
03 卧室 A / Bedroom A
04 卫生间 / Bathroom
05 卧室 B / Bedroom B
06 健身房 / GYM
07 卧室 C / Bedroom C
08 佣人房 / Maid's Room

整个项目采用了现代轻奢的材质和配色方案，以符合别墅主人低调奢华的生活品位。室内外墙体皆为中性灰色调，主要建材包括灰色大理石及墙布、榉木木皮、粉色系列家具及定制吊灯。量身打造的钢化玻璃和白色镭射切割扶手楼梯，各自为室内及室外的垂直动线提供了独特的空间体验，虚实、光影关系以不同手法产生。

THE FOCUS OF LIGHT LUXURY
轻奢聚焦

← 楼梯间垂直而下的吊灯，熠熠生辉，给空间增加适度的奢华。

一层平面图
1st floor plan

01 客厅 / Living Room
02 餐厅 / Dining Room
03 厨房 / Kitchen
04 卫生间 / Toliet
05 卧室 / Bedroom
06 车库 / Garage
07 阳台 / Balcony

独栋别墅拥有足够的空间享受个人生活，背靠山林，远离尘嚣。不被奢华所淹没而更加注重生活品质感的空间，关注的是修缮人与周围场域的关系，以达到和谐共处的生活状态。业主既能自由驱车去都市中心，也能随时抽身于市区的热闹，回归到这一方不被打扰的天地，与家人守候这片身心栖居的乐园。

轻奢生活
THE LIFE OF LIGHT LUXURY

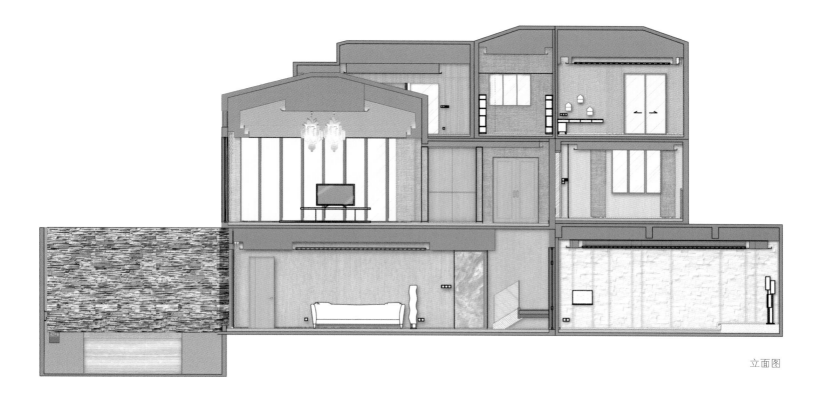

立面图

二层平面图
2nd floor plan

01 主卧 / Master Bedroom
02 女儿房 / Daughter's Room
03 卫生间 / Bathroom
04 客厅上空 / Over The Living Room
05 阳台 / Balcony
06 衣帽间 / Walk-in Closet
07 楼梯间 / Staircase
08 儿童区 & 阅读区 / Children's Area & Reading Area
09 采光井 / Light Wells
10 卧室 / Bedroom
11 主卫 / Master Bathroom

TO COMPANY YOUR FAMILIES AS LUXURIOUS AS POSSIBLE IN FANGYUAN

方圆之间，奢于亲人陪伴

LIGHT LUXURY STYLE IN HONGKONG

港式轻奢

● 项目信息　PROJECT INFORMATION

设计公司 / Millimeter Interior Design Limited
设计师 / Michael Liu
项目地点 / 澳门

项目面积 / 1022 m²
主要材料 / 玻璃、木地板、木材、石材等
摄影师 / Millimeter Interior Design Limited

扫码查看电子版

 # THE CONCEPT OF LIGHT LUXURY
轻奢理念

In this design, the designer uses geometric as the chief source, which runs through the outside and inside of the space. The designer uses the rectangle as the architectural appearance in particular which allows this whole house a sense of sporty and energy. Also, the designer uses many natural materials, such as wood and marble as its floor and wall and decorate the whole room with the grass wall to give out a sense of nature.

设计上，设计师以几何作为设计蓝本，贯穿外型及室内空间，并特意用矩形作建筑外形，弧形线条作室内空间。长方形的建筑外形与室内柔和的曲线形成鲜明对比，在整个房子里产生了一种充满活力的运动感。材质上则以自然物料作主调，包括使用木材、云石作墙身及地板，并配以草墙作点缀使得整个家散发着自然气息。

轻奢聚焦
THE FOCUS OF LIGHT LUXURY

功能上，设计师将原有楼梯位改到另一侧墙边，如此一来，除了将客厅空间扩大外，更能提高其他空间的使用率，不仅能为两个儿子、一个女儿提供房间，还能够为外婆提供宽敞的专属套房。而主人房独占一层，内设有空中花园、衣帽间、影音室等设施。

由于三个小朋友均在求学阶段，设计师特意将地下室改成与花园相连的阅读室，原有的花园因而被一道高墙阻隔，导致光线不能进入地下室。设计师特意将此高墙拆掉，改成为一道艺术装置的楼梯，使阳光顺利进入地下阅读时，并起到连贯上下两个花园的作用。地下室没有窗户的另一面则改成酒窖，内可摆放500支以上的藏酒。

THE LIFE OF LIGHT LUXURY
轻奢生活

老有所终，幼有所长，是古人"大道之行"的期盼，而今"陪伴"则变得弥足珍贵。当三代同堂，彼此拥有了"一碗汤的距离"，情感的丰满能让生活拥有生动的色彩，那是华丽物质无法企及的充实。

LIGHT LUXURY TAIWAN

○ 现代轻奢 / Modern Light Luxury

STYLE IN

台式轻奢

LIGHT LUXURY STYLE IN TAIWAN

轻奢理念 / The Concept of Light Luxury
轻奢聚焦 / The Focus of Light Luxury
轻奢生活 / The Life of Light Luxury

轻奢印象
THE IMPRESSION OF LIGHT LUXURY

　　台式轻奢在利用本土元素的基础上，充分吸收日式设计优点，创造独具台湾风味的轻奢华设计。此外，受到现代风格"突破传统，创造革新"的影响，越来越多台湾设计师主张"客制化"，将传统民风结合亚洲地方特色，以"人本主义"为出发点，尊重使用者需求，创造美学与实用并存的空间效果，形成新现代主义的台式轻奢风格。

Light luxury style in Taiwan is based on local elements, and thoroughly absorbing the advantages of Japanese design to creating a light luxurious with Taiwan style. In addition, influenced by the modern style of "Breaking through tradition and creating innovation" more and more Taiwanese designers advocate "Customisation" combining traditional folk with Asians' characteristics, taking "humanism" as the starting point, and respecting user's needs to creating a space that combines aesthetics and practicality, forming a new modernism light luxury in Taiwan style.

1 林政纬｜印象马来西亚
客厅

→ 1

01 轻奢理念　THE CONCEPT OF LIGHT LUXURY

台式轻奢结合了传统台式设计的简约格调，空间设计趋于极简，但不像极简风格那么冷淡、孤寂，利用不浮夸的装饰、精致完美的细节表达来传达现代人务实而雅致的审美品位。注重科学风水——"阳光""空气""水"——生命三元素的合理配比，关注建材的安全与健康，从而构建环境的人文舒适性。

It adopts simplicity of traditional Taiwanese design. The space design tends to be minimal, but not as cool and lonely as the minimalist style. It uses the unflattering decoration, exquisite and perfect details to convey the pragmatic and elegant aesthetic taste of modern people. Paying more attention to scientific Feng Shui — "Sunshine", "Air" and "Water" — the reasonable proportion of the three elements of life, and focusing on the safety and health of building materials so that it can build the human comfort of the environment.

02 轻奢空间　THE SPACE OF LIGHT LUXURY

讲究简约、工整、自然开阔的空间体验感，没有复杂的线条。空间气质克制而细腻，理智而优雅，其因地制宜的空间布局能将空间利用率提升到极致。

It pays attention to the simplicity, neatness, and natural open space experience, without complicated lines. Space temperament is restrained and delicate, rational and elegant, and its spatial layout can improve the space utilisation to the extreme.

03 轻奢色彩　THE COLOUR OF LIGHT LUXURY

空间配色沉稳内敛，多采用中性偏冷色调营造类似于日本"侘寂"文化谦逊、自然、朴素的艺术气氛。台式轻奢在传统台式设计常见的灰色、木色、深蓝色等基础上增加亮色、跳色，以玻璃、亚克力、不锈钢等新型材料衬托空间的轻奢质感。

Space colour is calm and restrained, and the cool neutral tone is used to create a modesty, natural and simple artistic atmosphere which is similar to the Japanese "Wabi-Sabi" culture. The Taiwan Affordable Luxury based on the familiar tones such as grey, wood and dark blue in traditional Taiwanese design. It adds bright colour and colour jump, using glass, acrylic, stainless steel and other new materials to set off the luxurious texture of the space.

04 家具配饰　FURNITURE ACCESSORIES

坚持精工品质，匠人精神，引进智能家电，强调功能并追求家具的优良触感，没有赘余的装饰；清晰明朗的家具布置格局描绘出克制与理性、雅致而舒适的空间气质。

It adheres to the quality of craftsmanship and carries forward the spirit of the craftsman, The introduction of smart home appliances, emphasises functions and pursues the excellent touch of furniture without any surplus decorations. The clear furniture layout patterns depict restraint and rational, elegant and comfortable space temperament.

2　马健凯｜葛里法钱宅
电梯玄关区

3　林政纬｜印象马来西亚
餐厅

4　刘荣禄、黄沂腾、陈福南｜泉州卡飒明筑 A 户型
楼梯

THE GRAND TROPICAL HOUSE AND VACATION SENTIMENT

热带住宅的恢弘与度假情调

LIGHT LUXURY STYLE IN TAIWAN

台式轻奢

● 项目信息　PROJECT INFORMATION

项目名称 / 印象马来西亚　　　项目地点 / 马来西亚
设计公司 / 大雄设计　　　　　项目面积 / 398 m²
设计师 / 林政纬　　　　　　　主要材料 / 大理石、磁砖、特殊玻璃、镀钛、特殊漆等

扫码查看电子版

The detached villa is located in a place where every day is summer. This villa is around with a courtyard, and the designer fully uses this advantage to decorate this space. The owner wants the villa to become a light luxury villa with a magnificent atmosphere and can be a place that has both private and public areas. After surveying the house entirely, the designer decides to create this dream house with the impression of the second nature and life.

这座独栋别墅所在的地区四季如夏,自带四周庭院环绕的优势,因而设计师在做空间策略时对这个特色进行了呼应。房子的主人希望住所要有轻奢、恢弘大气之势,且对家人的共同活动空间与私人空间有强烈的分享与隐私要求,由此设计师在深入勘察住宅情况之后,以创造第二自然、带入生活印象为方针,试图打造业主心中梦想的家。

THE CONCEPT OF LIGHT LUXURY
轻奢理念

轻奢聚焦
THE FOCUS OF LIGHT LUXURY

通过加长电视主墙,迎宾玄关有了完整而大气的回廊切面,对户外和进门的宾客形成一定的视觉屏障,增强室内的私密性。在回廊结构中,大理石材质和灯光打造出的质感给宾客留下关于这个家的第一印象,起到装饰性的作用。

保有阳光、空气、微风串流的动线规划是设计师的首要目标。开放的餐厨空间、客厅等形成一个环环相扣的通透格局,让阳光、风和空气自由来往。餐桌、中岛台、厨房操作台一字排开,而浅色桌面、台面与灰调的拼贴地面、椅子搭配协调,营造出一个理性、大度的用餐、烹饪区域。独特的酒窖设计凸显了主人的生活品位,可供一家人享用,也可接待宾客,还与柜体、置物架一起强化了这部分空间的收纳功能。

餐厅轴线的一侧是开阔的降板客厅。降板的低调设计,让视线连接窗外的发呆亭与绿意。而室内不同的材质、纹理以及抬高的沙发也构成客厅的场域包围线,且视野开阔,会客、谈天与赏景皆适宜。电视背景墙中,强调自然的木质格栅衬托粗犷的石材,给人一股冷静与暖流兼具之感。

二层是主卧、两个孩子的卧室以及长辈房,每个卧室都添设了收纳空间,方便临时储物或是随时取阅图书。此外主卧还配备了书房和男女主人各自的更衣室,而书柜设计与工字拼地板呼应,进一步突出室内明亮、静谧的特点。

→ 木料、镜面材质与照明共生出低调的特性和高级感

↓ 木格栅赋予空间细腻、自然的观感,常有隔而不断的效果

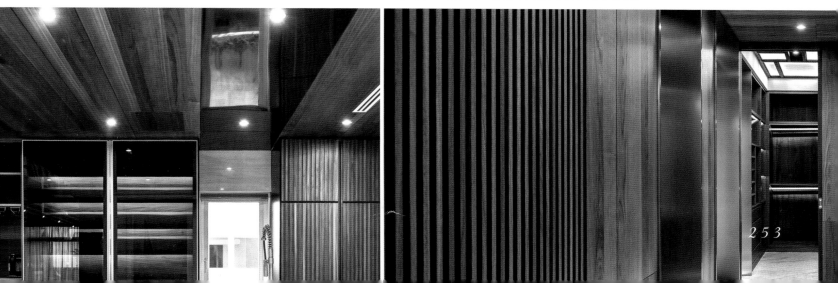

THE LIFE OF LIGHT LUXURY
轻奢生活

轻奢生活的现代意义在于品位和品质，设计师运用照明设计呈现简洁干练、线条感、秩序感的效果，以大小不一的开窗满足通风与采光需求，使用十多种不同纹理的瓷砖与石材、金属、木皮等交互搭配，并通过现代语汇表现以及诠释当代空间，在热带住宅中增添舒适的清凉感，同时不失奢华与度假风情。

→ 以玻璃作为书房与其他空间的隔断，结合大理石地面与LED灯条之间的映射，给人通透、敞亮的感觉。

The Amazing House of Retired Enterpriser

退休企业家的魅力宅邸

LIGHT LUXURY STYLE IN TAIWAN

台式轻奢

● 项目信息　PROJECT INFORMATION

项目名称 / 葛里法钱宅
设计公司 / 界阳 & 大司室内设计有限公司
设计师 / 马健凯
项目地点 / 台湾桃园
项目面积 / 340 m²
主要材料 / 镀钛、锈化镜、石皮、钢刷木皮、特殊漆等
摄影师 / Yana Zhezhela, Alek Vatagin

扫码查看电子版

The outside atmosphere of this house is based on the professional layout, and the inner atmosphere is based on its exquisite detail design. Faced with this one floor one luxurious apartment, the designer designs this house based on the couples daily lives, and uses the opening-up and cooperation way to divide the cross transparent public space with the great pattern from inside to outside. Stepping out the elevator, people can feel a natural and smooth style, gold and black barriers are staggering which allow the light to penetrate and emit freely. The site planning shows the concept of humanity, and the detail indicates the information of space which builds the unique temperament and style of this house.

THE CONCEPT OF LIGHT LUXURY
轻奢理念

宅邸的气度源于专业布局，而室内的氛围则来自于细腻的细节设计。面对作为企业家退休后乐活宅邸的一层一户大平层空间，设计师以夫妻二人的起居动线为依据，以开放融合的方式对横向通透的公领域空间进行区块划分，将大气的格局从室内延伸到室外。踏出电梯，则迎来行云流水的风格，金与黑交错的隔栅排列让光线能恣意地穿透、流泻；场域的规划表现抽象的人文概念，细节的刻画孕育出空间底蕴，形塑出属于这套房子特有的气质与风情。

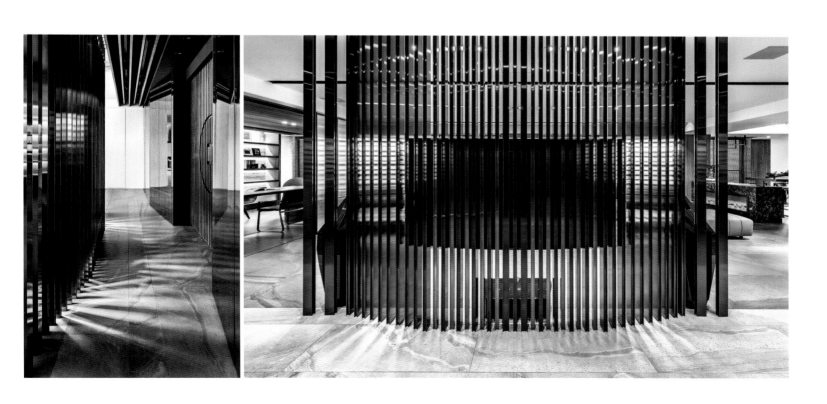

THE FOCUS OF LIGHT LUXURY
轻奢聚焦

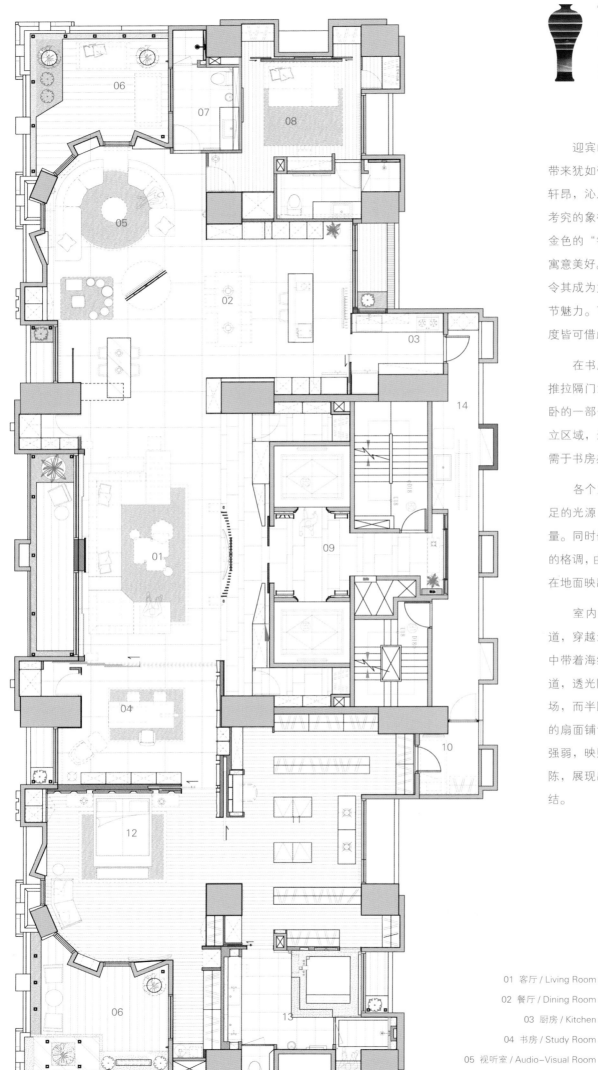

迎宾的气势在玄关的对接中展开序幕，带来犹如帝苑级的格局礼赞，稳重、壮观、轩昂，沁入震撼。双开的电动门上，雕琢着考究的象征尊贵的中国帝制纹理，中间刻有金色的"钱"字样，而两个半圆合为一圆也寓意美好。以这样的方式标注出屋主的姓氏，令其成为如家徽般的象征，突出了设计的细节魅力。可旋转的大荧幕格外实用，任何角度皆可借此观赏娱乐。

在书房与客厅的分界中，设计师以大面推拉隔门为隔断。隔门敞开时，书房即为主卧的一部分；隔门拉上时，书房则界定为独立区域，这种设计更方便居住者在休憩前尚需于书房办公的隐私用途。

各个空间中的大开窗和小开窗提供了充足的光源，而多样的窗帘很好地控制了进光量。同时嵌入式灯具和简约的吊灯契合空间的格调，由此自然光或人工光透过多种材质，在地面映出迷人的光影，令人沉醉。

室内的视觉动线如步过层层回旋的过道，穿越无数楼阁与屏栅，而后迎来的气势中带着海纳百川的度量。走进双通的回向廊道，透光隔屏间投射出的光线如挥剑般有气场，而半圆弧的隔栅引导光质，其影如展开的扇面铺设。光影的展示与演变随日照变换强弱，映照在纹理清晰的灰石砖上，脉络铺陈，展现出空间里自然力与材质间的交融连结。

平面图 / SPACE PLAN

01 客厅 / Living Room
02 餐厅 / Dining Room
03 厨房 / Kitchen
04 书房 / Study Room
05 视听室 / Audio-Visual Room
06 露台 / Terrace
07 客卫 / Guest Bathroom
08 客房 / Guest Bedroom
09 电梯玄关区 / Elevator Entrance
10 佣人房 / Maid's Room
11 衣帽间 / Walk-in Closet
12 主卧 / Master Bedroom
13 主卫 / Master Bathroom
14 阳台 / Balcony

THE LIFE OF LIGHT LUXURY

——— 轻奢生活

　　镀钛的使用带来色泽艳丽的效果，锈化镜虽显斑驳但其金属色更为突出，而木质和石材的运用使得空间氛围趋向从容，因而呈现眼前的便是轻奢但不肤浅、稳重却不刻板的雍容典雅的大宅风范。居于此，行走坐卧之间，不知不觉便会被熏陶得更自在，感悟生活的所赠所予。

265

THE PERFORMANCE OF GORGEOUS AND ABSTRACTION

华丽与抽象的演绎

LIGHT LUXURY STYLE IN TAIWAN
台式轻奢

● 项目信息　PROJECT INFORMATION

项目名称 / 泉州卡飒明筑 A 户型
设计公司 / 超美学事业体·刘荣禄国际空间设计
主要设计师 / 刘荣禄、黄沂腾、陈福南
协同设计师 / 周筱婕、邱如怡、谢宜臻
项目地点 / 福建泉州

项目面积 / 550 m²
主要材料 / 镀钛钢板、钢板烤漆、绷皮、镜面、白膜玻璃、茶玻、圣罗兰黑金石材、松柏石、黑金石、灰貂、卡拉拉白大理石、黑檀木木皮、橡木木皮、刷漆、壁纸等
摄影师 / 李国民影像事务所

扫码查看电子版

Designers do not focus on any existing design styles or concepts; they are beyond restrictions and are expected to the futurity of the spirit and aesthetic, devote particular care to the ground-breaking thinking based on customers' demand, and hold "Design makes art life" idea to realise the philosophical concept and the aesthetic of life. In the premise of use efficiency and space beautifying, designers provide a specific expectation for the indoor area so that they can create a future space and start a new imagination of life.

After defining "the performance of gorgeous and abstraction" concept, designers emphasise on using and collocating various kinds of materials to realise the space temperament with gorgeous and fashion, especially expressing the affordable luxury concept, and perform and exhibit the higher taste.

THE CONCEPT OF LIGHT LUXURY
轻奢理念

　　刘荣禄先生不拘泥于任何既有的设计风格及理念，以精神与美学的未来性为期许，讲究以客户需求为基底的开创性思维，秉"艺术向往生活"为哲学理念与生活美学。由此，其在使用效率与空间美化两者兼备的前提下，提出对室内空间的具体未来展望，从而为众人创造未来空间并开创崭新的生活想象。

　　在将空间定义为华丽与抽象的碰撞之后，刘荣禄先生通过各种各样材质的运用与搭配，呈现华丽与时尚的空间气质，尤其表现处于华丽表面之下的轻度奢侈，演绎并展现极高的品位。

 # THE FOCUS OF LIGHT LUXURY
轻奢聚焦

富有几何造型立体变化的天花板、大面积的黑色抛光大理石地面分别占据了天与地，此时充足的日光从屋外迎面而来，在视线所及之处将华贵兼具时尚的气韵展示无余，使空间绽放出充满智慧的品位。

多个不同几何造型切割交织所组成的电视墙体、墙面的垂直水平线条对应天花交错的几何量体、大理石材质的天然线条，于隐约之中将主客厅融入具有美术性的荷兰风格派的语汇里。屋内耸立的现代雕塑使屋主在返家进入客厅时，总会如亲身阅览影格般，欣赏这一个魔幻、迷人的空间。地面上富有质感的大理石搭配剔透的水晶吊灯，深邃却又明亮，勾勒出电视墙利落的线条，延伸着华丽与时尚的墙面，营造出大宅气势的底蕴。

墙面的石材营造出镜面效果，与光源共生出客厅的弘大气场。

THE LIFE OF LIGHT LUXURY
轻奢生活

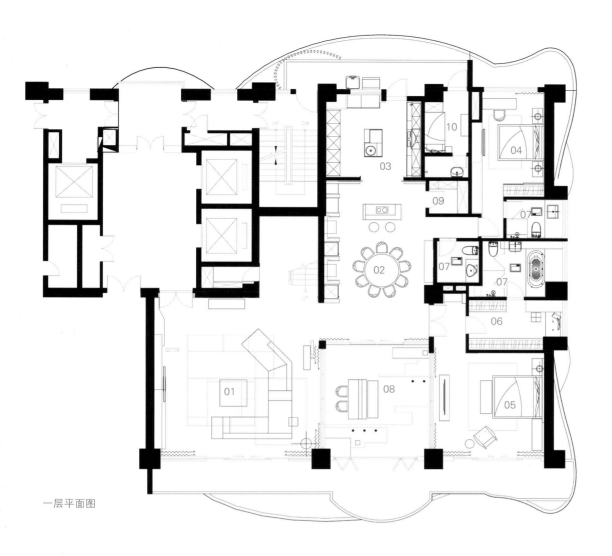

一层平面图

挑空层、楼梯间、途经餐厅通往室内花园、卧室、衣帽间，甚至连卫生间都体现了房子的配套齐全，且干练、简约，不损高贵与雅致。一石、一木、一金属，都诠释了节制与高雅的居家风度。

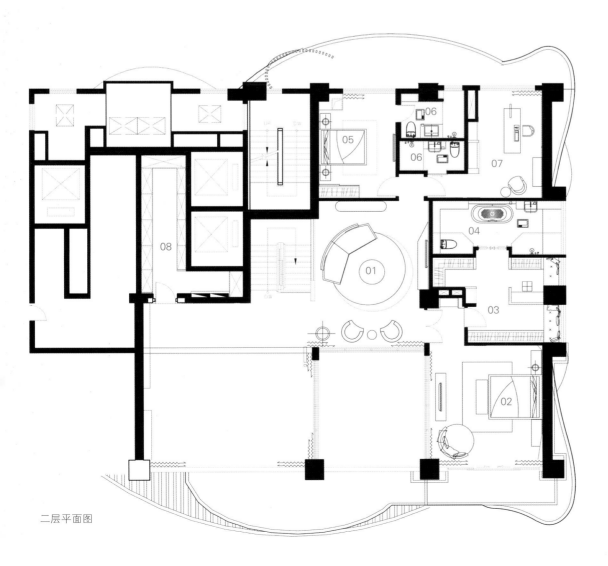

二层平面图

一层平面图
1st floor plan

01 客厅 / Living Room
02 餐厅 / Dining Room
03 厨房 / Kitchen
04 卧室 A / Bedroom A
05 长亲房 / Elder's Bedroom
06 衣帽间 / Walk-in Closet
07 卫生间 / Bathroom
08 室内花园 / Garden
09 储藏室 / Storage
10 佣人房 / Maid's Room

二层平面图
2nd floor plan

01 起居室 / Family Room
02 主卧 / Master Bedroom
03 衣帽间 / Walk-in Closet
04 主卫 / Master Bathroom
05 卧室 B / Bedroom B
06 卫生间 / Bathroom
07 书房 / Study Room
08 储藏室 / Storage

STYLE FURNITURE PERFORMS PRELUDE TO LIFE

格调家居演绎生活序曲

台式轻奢 LIGHT LUXURY STYLE IN TAIWAN

● 项目信息　PROJECT INFORMATION

项目名称 / 生活序曲 -1680
设计公司 / 格伦设计工程
设计师 / 虞国纶
项目地点 / 台湾台北

项目面积 / 215 m²
主要材料 / 水染木皮、棕榈灰大理石、雕刻黑大理石、订制铁件、陶瓷烤漆、进口瓷砖、订制植生墙、原装家具、原装灯具等
摄影师 / Ken

扫码查看电子版

 THE CONCEPT OF
LIGHT LUXURY
轻奢理念

This design starts from the roughest to create a unique space and function allocation. The designer uses the extension of the visual, the overlapping of space and unevenness materials to create a public domain with decent, a master room with rank and honour and a girl's room with a delicacy. This design solvates the best space use and at the same time allows people to have an excellent affordable luxury feeling. Music, arts and literacy are in close contact with each other which bring lively rhythm in that space, and the spearhead of great momentum and restraining brings the vigorous and firm. The concise dynamic, the measurement, the extension of space shaping and copy space with overlapping and moderate, all of these formed a perfect house.

本案由毛胚阶段开始着手规划，从零开始打造出专属的使用空间与机能配置，运用视觉的延伸、空间的重叠及高度参差的材质分界打造出纵深大器的公共领域、尺度尊荣的大主卧与精致女孩房等，媒合最佳的空间使用率与非凡情境的轻奢感受。音乐、艺术与书香的亲密交融，让空间从此有了鲜活的生命律动，磅礴却又内敛的锋芒带出雄浑气势，精炼的动线、量体、空间造型的延伸、交叠以及适度的留白，成就了无与伦比的心灵美宅。

THE FOCUS OF LIGHT LUXURY

○———— 轻奢聚焦

　　入门玄关区映入眼帘的是两侧迤逦开展的白色立体序列，高低起伏的30°锥体，极其精致。顶部在洗墙光源映照下，呈现明暗有致的光影层次。其设计灵感来源于教堂内巨大的管风琴，意象隐喻满室悠扬的复调旋律，动静之间，由此处拉开生活的序幕。空间配色沉稳大气，既有亲近自然的悠然宁静，又有身处都市的时尚品位。石材的中性冷静质感与木色基调的自然温润，邂逅出空间低调的奢华属性。灰色给予空间平和氛围，清凉的蓝灰色休闲而精致，是客厅中最明丽的那抹风景。

THE LIFE OF LIGHT LUXURY

轻奢生活

家是一个让人身心皆可轻松安放的地方，就像本案所呈现出的那样，为业主提供属于一家三口的幸福时光。空间注重私人领域的设计，女儿房同样以套间形式规划，留给孩子更多个人独立支配的空间。这种相互依赖又相互独立的空间布局，正是父母对孩子最好的爱——既给你足够的空间成长，又在你需要的时候陪伴在你身边。轻奢生活便是如此，让家中的每个人都能享受到更有品质感的生活。

THE CONCEPT OF LIGHT LUXURY
轻奢理念

This house is an old double deck with 20 years age of the property in the centre of Taipei, and It is six meters high and has its private garden. After examining on the spot, the interior arrangement is jam and closed which cause people who live there cannot see the sunlight often and cannot breathe fresh air. If the existing structure traps the designer, none of the solutions can be used, so it is evident that the designer has to find a new path.

The designer uses enlarging space affection as advanced to transfer the user's real needs and functions into a rationalised allocation and aesthetic arrangement, which allows the final design looks more spacious and suitable. Specifically, with the permission of the management of that building, the designer starts to remove compartment, enlarge the window and reconsolidate the pattern and inside-and-outside relation to allow more sunlight, views and wind in the house without reflecting the structure safety which benefits both the old building and nature. This design represents a tone of black and white, the fashion and modern avant-garde style.

这是位于台北市中心的已有20年屋龄的老旧楼中楼，拥有6m挑高及私人庭院。实地勘察发现，其内部格局堵塞、封闭，导致日光及气流难以通行，如果困于框架中打转，任何方案都将窒碍难行，所以势必得另辟蹊径。

设计师以放大空间效果为前提，将空间使用者的实际需求及机能落实为合理化的配置及美学安排，使完工后的作品呈现更为宽敞大器、舒适快意的理想居宅样貌。具体而言，经大楼管委会许可，在不影响结构安全的情况下，设计师着手移除隔间、大开窗户，重新整顿格局和内外关系，让日光、竹景、微风穿窗而入，让旧楼得以与自然相存相惠，呈现出黑白、时尚、前卫的现代住宅格调。

THE FOCUS OF LIGHT LUXURY
轻奢聚焦

水平建构

空间价值能否发挥，取决于平面关系的建立。设计师使客厅、餐厅、吧台、厨房呈一字符串联，贴近窗外的光和景，达到空间层次拓宽及深化的效果，仿佛里外没了隔阂，编排出敞亮且互动频密的场域。

垂直贯连

相较于没由来地扩充面积，设计师重视纵轴线的经营，局部消解二楼面积，归还给客厅最大尺度；在掣引所有日照、庭院动态之余，通过金属吊灯作为地面及空中走廊的共享元素，串联上下关系，也呈现出 6m 高的磅礴气势。

形质之美

为达简而美的诉求，避免材料铺张，仅通过对比手法拾掇形质之美，如粗犷与细腻、时尚和温润，皆延伸于线面的虚实起伏间；光照映耀下，自然缔构肌理与明暗变化，空间也因时序变迁产生独特丰韵。

1 多元运用：不同材质与色彩的混合搭配，产生个性化的几何美。
2 活用角落：楼梯采用镂空设计，并在下面的小空间里布置石子、木块、绿植、艺术画，具有宁静、惬意的造景效果。

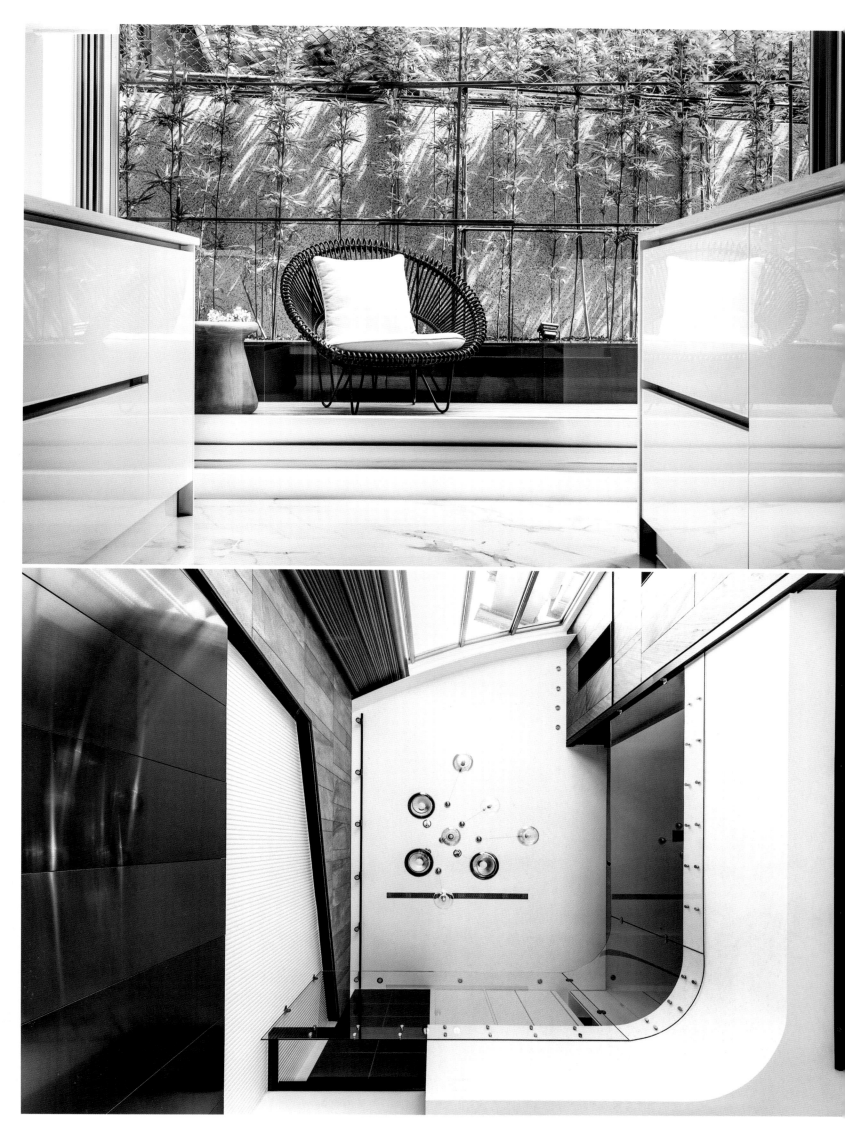

THE LIFE OF LIGHT LUXURY
轻奢生活

设计师以白色为空间底色，以艺术品、灯光布置以及亮色为生活增温。在线面的虚实起伏间，强调一种收敛的、有节制的时尚前卫感，让人获得一股仿佛走入纽约博物馆般的艺术韵致。而只要敞开折门，目光即可与两方庭院景致相接，让空间浸润户外的惬意，心亦随之松弛，慵慢自得，享受生活中的日常意趣。

THE DENSE INKED FLAVOUR BUILDING A SPACE WITH TEMPERATURE

筑有温度的低奢场域

LIGHT LUXURY STYLE IN TAIWAN

台式轻奢

● 项目信息　PROJECT INFORMATION

项目名称 / 墨韵岑砌
设计公司 / 维耕设计
设计师 / 林志龙、苏楠凯
项目地点 / 台湾高雄
项目面积 / 479m²
主要材料 / 实木皮、石材、镀钛铁件、黑镜、壁布等
摄影师 / 林明杰

扫码查看电子版

THE CONCEPT OF LIGHT LUXURY
轻奢理念

The designer believes that design is the attitude of personal life and the extension and reflection of aesthetic style, so the designer should pay more attention to the place spirit. In this house, this design focuses on smartness and based on the specific situation, the designer uses many features such as the dynamic, the perception, the texture and the ideological and practical work to achieve oneness so that it can combine the function and sense of beauty through decorations and build this house with ink edification, low-profile and luxury temperature.

  设计师认为，设计是个人生活态度和美学风格的延伸与体现，应着重于空间设计上的场所精神。在这套房子中，设计师以现代风格为主调，根据案例具体情况，思考动线、比例、肌理与虚实等特点，进而从一体性概念出发，通过对装饰艺术的运用与呈现，塑造机能与美感的完美融合，为屋主打造墨韵熏陶的、发自内心的低调与奢华气场。

THE FOCUS OF LIGHT LUXURY
轻奢聚焦

客厅之中，目光所及是风格独具的软饰搭配，灰阶充当主角，深蓝化身为辅佐，构筑出优雅和谐的场景。一旁，设计师运用落地窗带来采光及景观，呼应着室内植栽摆饰，造就生动的生活画面，并让大理石立面串联领域，从客厅一路平整延伸至餐厅，转换成展示墙背景，点出墙面的雅韵风度。

餐区以一座中岛增强机能，可储物收纳，可简易操作，搭衬飘逸自然的花艺陈设以及极具质感的吊灯布置，落实品位十足的用餐氛围，让居住者在餐厨空间里更自如方便。

卧室内，设计师除了赋予其偌大的空间感，更配置书桌、起居、更衣间等饭店式机能，再加入层次分明的场域布局，使视觉可一路与户外阳台连贯，划出充满延伸性的寝居轴线。卧室的起居地带，透过充满戏剧张力的壁材表现，带出行云流水的泼墨纹理，瞬间让墙面变得精彩不凡，达到转换空间气度的作用。

一层平面图
1st floor plan

01 客厅 / Living Room
02 餐厅 / Dining Room
03 厨房 / Kitchen
04 卫生间 / Bathroom
05 阳台 / Balcony

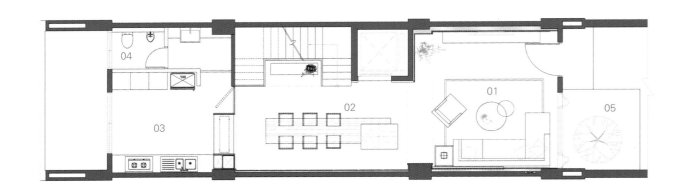

二层平面图
2nd floor plan

01 卧室 A / Bedroom A
02 卫生间 / Bathroom
03 楼梯间 / Staircase
04 衣帽间 / Walk-in Closet
05 阳台 / Balcony

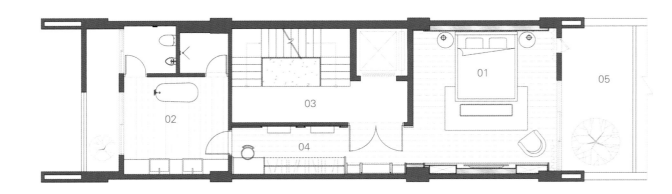

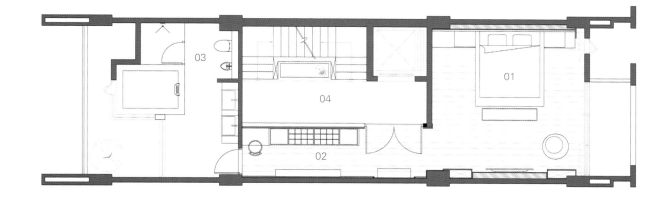

三层平面图
3rd floor plan

01 卧室 B / Bedroom B
02 衣帽间 / Walk-in Closet
03 卫生间 / Bathroom
04 楼梯间 / Staircase

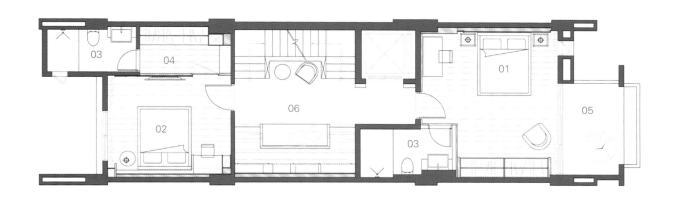

四层平面图
4th floor plan

01 卧室 C / Bedroom C
02 卧室 D / Bedroom D
03 卫生间 / Bathroom
04 衣帽间 / Walk-in Closet
05 阳台 / Balcony
06 起居室 / Family Room

THE LIFE OF LIGHT LUXURY
轻奢生活

一切空间规划，都应因人而存在。设计师运用大件的建材铺陈出低奢的气场，同时加入绿植、家居摆件等塑造场域温度，且让每项机能都贴近屋主一家的生活轨迹，打造出奢华、温馨兼具的住宅视野。

图书在版编目（ＣＩＰ）数据

现代轻奢：当代住宅设计 / 深圳视界文化传播有限公司编．-- 北京：中国林业出版社，2018.10
ISBN 978-7-5038-9768-9

Ⅰ．①现… Ⅱ．①深… Ⅲ．①住宅－室内装饰设计 Ⅳ．①TU241

中国版本图书馆CIP数据核字（2018）第228730号

编委会成员名单
策划制作：深圳视界文化传播有限公司（www.dvip-sz.com）
总 策 划：万　晶
编　　辑：杨珍琼
校　　对：陈劳平　尹丽斯
翻　　译：马　靖
装帧设计：叶一斌
联系电话：0755-82834960

中国林业出版社 · 建筑分社
策　　划：纪　亮
责任编辑：纪　亮　王思源

出版：中国林业出版社
（100009 北京西城区德内大街刘海胡同 7 号）
http://lycb.forestry.gov.cn/
电话：（010）8314 3518
发行：中国林业出版社
印刷：深圳市雅仕达印务有限公司
版次：2018 年 10 月第 1 版
印次：2018 年 10 月第 1 次
开本：235mm×335mm，1/16
印张：20
字数：300 千字
定价：428.00 元（USD 86.00）